DREAM SCIENCE

DREAM SCIENCE

EXPLORING THE FORMS
OF CONSCIOUSNESS

J. F. PAGEL

AMSTERDAM • BOSTON • HEIDELBERG • LONDON • NEW YORK •
OXFORD • PARIS • SAN DIEGO • SAN FRANCISCO •
SINGAPORE • SYDNEY • TOKYO

Academic press is an imprint of Elsevier

Academic Press is an imprint of Elsevier
The Boulevard, Langford Lane, Kidlington, Oxford, OX5 1GB, UK
225 Wyman Street, Waltham, MA 02451, USA

First published 2014

British Library Cataloguing-in-Publication Data
A catalogue record for this book is available from the British Library

Library of Congress Cataloging-in-Publication Data
A catalog record for this book is available from the Library of Congress

ISBN: 978-0-12-404648-1

For information on all Academic Press publications
visit our website at **store.elsevier.com**

Printed and bound in the United States

12 13 14 15   10 9 8 7 6 5 4 3 2 1

 Working together
to grow libraries in
developing countries

www.elsevier.com • www.bookaid.org

Dedication

Dedicated to Bert (States) & Ernie (Hartmann)
Sorely missed free spirits of dream science

Contents

Acknowledgments

More than 40 years ago, working as a young neurochemist, I found my way into the developing science of dreaming with the help of Vernon Pegram, then director of neuroscience at the University of Alabama. In the intervening years leading to this book there have been many who have helped in the journey, and I would like to personally acknowledge the help and support of Earnest Hartmann, Bill Domhoff, and my co-authors on dream-based projects: Barbara Vann (Chair, Department of Sociology, Loyola Baltimore) Carol Kwiatkowski (Epidemiology: CU – Boulder) and Kathleen Broyles (Santa Fe College of Art and Design, Millagro at Los Luceros, Sundance). For this project, I had specific support from Joan Tewkesbury (Screenwriting Advisor, Sundance Labs) and Richard Moore (Chair, Humanities (retired), University of Alabama in Huntsville) in straightening the original path, Henry Cleary (Professor Emeritus Neurology, UCol.HSC) and Janet MacKenzie (Executive Director of the Mesa Prieta Petroglyph Project) in addressing anthropology, Reyes Garcia (Professor Emeritus Philosophy, Ft. Lewis College) and Susan O'Leary in reviewing the included philosophy, meditation, and semiotics. This project would not have been possible without Kathleen Broyles' intellectual and editorial efforts as well as her love and support. The cover image has Barbara Zaring's beautiful work "Stardust" as its background. In the foreground is my own calligraphy, the Chinese glyph for "dream," developed under the tutelage of the master calligrapher: Kazuaki Tanahashi.

About the Author

J.F. Pagel MS/MD is an Associate Clinical Professor at the University of Colorado School of Medicine and Director of Sleep Disorders Center of Southern Colorado. He has served as chair of the Education Committee for the American Academy of Sleep Medicine, on the Sleep Medicine Board Testing and Policy Committee for the American Board of Internal Medicine, and on the Board of the International Association for the Study of Dreams. His published work includes >150 articles, and co-authorship of the initial Sleep Medicine diagnostic code description for Nightmare Disorder. His books include: Primary Care Sleep Disorders – A Practical Guide, Humana Press, (co-ed. 2007), The Limits of Dream – A Scientific Exploration of the Mind/Brain Interface, Academic Press (2008), and Dreaming and Nightmares, Saunders (ed. 2010).

Preface: The REMS Equals Dreaming Debacle

A dream is a story with something missing. **(Pagel; adapted from Hunt, 1991) (1)**

What we call a dream differs from person to person. Dreaming, from multiple perspectives, has led the way to the creation of science – in its most basic form, the logical organization of consciousness. A dream is elusive and has no external markers. Yet science has taken much from dreams.

The cave paintings of southern Europe, the earliest markers for the onset of modern consciousness, strongly reflect an imprint of dreams. Dreams illuminated an understanding of interior and external reality that theretofore had evidently not been seen. Present at the Paleolithic onset of our species' consciousness, dreams have persisted as a creative source for art, philosophy, and science. The pedigree of scientists and philosophers delving head-first into dream-based theory is astounding. They include the developer of the scientific method, the founders of psychoanalysis, and the discoverers of the DNA helix. Through dreams, a host of theorists purport to have discovered the basis of mind and consciousness. And yet, dreams remain poorly defined. Dreams are personal and private. To study an actual dream is to study its description. Dream sharing is a complex and unsettled area of social discourse, affected and altered by interaction, and easily manipulated by the observer and the methods of study. Yet, dreaming has an incredibly long track record of study. We have been interpreting dreams and their meanings for more than 4000 years.

In today's world, both the scientist and the philosopher typically ignore the significance of the dream as an instrument for scientific understanding. They have most often chosen to accept simple answers about dreaming without questioning, an approach with both individual and epistemological risk. The dreamer who does not acknowledge dreaming derives little from dreams. The scientist who accepts unsubstantiated theories of dreaming as truth is accepting a distorted view of scientific reality. Dreams are at the basis of the major theories of mind, philosophy, and brain function. Dreams may actually be important.

URBAN MYTHS

Early in the twentieth century, Sigmund Freud was among the first modern scientists to emphasize the potential cognitive importance of dreams. His attempts utilized the dream as a window into understanding and analyzing the problems and processes of psychiatric illness (2). At the basis of his psychoanalytic theory was the concept of interpretable dream content and shared dream structure describing the functioning of the mind. Freud proposed that psychic structures and their dynamics could be inferred from the psychoanalytic interpretation of dreams. This information could then be used to develop a treatment plan for psychiatric symptoms. Psychoanalysis extended the scientific method of reproducible experiment and therapeutic outcomes into the investigation of the mind. As Freud stated, "Psychoanalysis is related to psychiatry approximately as histology is related to anatomy," (3) and "The study of dreams is not only the best preparation for the study of the neuroses, but dreams are themselves a neurotic symptom, which, moreover, offers us the priceless advantage of occurring in all healthy people" (4). For more than a generation, psychoanalysts used dream interpretation to make diagnoses and form treatment plans. In an era in which there were few alternative treatments available for psychiatric illness, psychoanalysis became the treatment of choice. Unfortunately, the psychoanalytic era of psychiatry turned into one of long-term institutionalized therapy. Today, other psychodynamic and medical approaches to the treatment of these illnesses have proven far more effective and much less costly. With the loss of the basic underpinning of therapeutic treatment, it would not have been surprising if psychoanalytic theories had collapsed, and fallen into disrepute. The outcome has been quite different. During the same period in which applied psychoanalysis seriously declined as a method for treating illness, it became one of the primary techniques utilized in attempting to understand the structure and function of the mind. The psychoanalytic perspective moved to fields outside psychiatry that were focused on aspects of higher cognitive functioning. Today, psychoanalytic constructs of mind are commonly used in attempts to understand how the structure of the mind might apply to the associative thought of creative process, the bidirectionality of cinema, the impulsive power of art, and the neurobiology of consciousness (5).

Freud had a brass plaque inscribed to hang over the desk of his study, stating, "On this site, Sigmund Freud discovered the meaning of dreams." According to an urban myth, his apartment was torn down to make room for a new highway. On a stone block in the center of a traffic interchange in downtown Vienna, you can view the plaque and potentially re-experience Freud's simple explanation for the meaning of dreams. Psychoanalysis is an important marker in our attempts to understand the functioning portions of the brain that we call "mind." It is different from other neuroconsciousness theories in its attempt to offer insight into the basic psychodynamics of psychiatric disease, and through the analysis of associative thought and dreams derive information as to the structural functioning of the brain.

SPONTANEOUS ELECTRICAL ACTIVITY

In the late 1920s, the German neurologist Hans Berger recorded spontaneous electrical activity from scalp electrodes. He was the first to describe the alpha rhythm, the predominant frequency-based electrical rhythm of the human central nervous system (CNS), a rhythm that is particularly apparent at the transition from drowsy wakefulness to sleep (6). In 1952, Eugene Aserinisky, a young graduate student at the University of Chicago, recorded rapid conjugate eye movements (REMs) during sleep. These REMs were initially thought to be noise in the sleep recordings. Another young graduate student by the name of William Dement developed multilevel monitoring into the first full-night polysomnographies. Every subject that he studied had episodes of REMs occurring in cycles approximately every ninety minutes during the night. When Dement awakened test subjects during these REM states, many reported having dreamed (7). This finding became the psychoanalytic "smoking gun" demonstrating that a biological substrate existed that was equivalent to dreaming. It seemed obvious that rapid eye movement sleep (REMS) was the dream state – evidence of Freud's structural mind child, the mythical "id" present in brainstem biological activity. The Freudians were back.

CLEAR AND IRREFUTABLE EVIDENCE

The news that neuroscientists had discovered a biological brain state equivalent to dreaming was even bigger news for philosophers. For over 500 years, philosophers had been trying to solve the problem of the relationship of mind to brain/body. Ever since Descartes, the Cartesian dichotomy between mind and brain had been codified and built into the structure of knowledge. The subjective was divided from the objective, conscious thought from unconscious, medicine from psychiatry, and science from art. But REMS was discovered – the apparent mind-based cognitive state of dreaming – clear and "irrefutable" evidence that body and mind were one and the same. Since that point it has been generally accepted that what we call the mind is the functioning of underlying brain activity. Today, it is difficult to realize the compelling power of the

Cartesian argument that body and mind are different kinds of stuff that somehow interact in the brain (8).

NEUROCONSCIOUSNESS

The discovery of REMS radically changed neuroscience. In the 1970s, the theory of activation synthesis postulated that all cognitive behaviors, both conscious and non-conscious, reflected the biological and physiological activity occurring in the brain (9). The primary proof for this theory was the apparent finding that REMS was the CNS dreaming state. The authors of activation-synthesis proposed that the cognitive activity of dreaming was based on CNS activation occurring during REMS. This postulate remains at the basis of current neuroscientific theories of consciousness including activation synthesis and the derivative offspring: activation, input, modulation (AIM), reverse learning, neural net theory, search-attention theory, and most recently protoconsciousness theory (10,11). These theories are based on the postulate that dreaming is equivalent to REMS. This primitive electrophysiological state of activation is integrated with upper cerebral cognitive processes to create dreams. Neuroconsciousness theories have been extended to levels of exceeding complexity. If REMS is dreaming, animal models and brain scanning studies of REMS must describe the cognitive state of dreaming. Using brain-slice studies, functional magnetic resonance imaging (fMRI), positron emission tomography (PET), magnetoencephalography (MEG) recorded from superconducting quantum interface devices (SQUID), and micropipette techniques, neuroscientists have experimentally described the neuroanatomical, neurochemical, electrophysiological, and neuropsychiatric characteristics of REMS. Most of this work is presented as dream research.

THE SPECIAL RELATIONSHIP

From the first polysomnography (PSG) studies, it was clear that REMS and dreaming were doubly dissociable (12). REMS occurred without dreaming, and dreaming occurred without REMS (13). It is still unclear whether any special relationship exists between REMS and dreaming (14). Yet there are entire fields of study and many famous scientists, philosophers, and physicians with considerable investment in the belief that REMS equals dreaming. Many deny that dreaming occurs throughout sleep, despite the overwhelming evidence that REMS is but one of the electrophysiological brain states associated with dreaming (15,16). Without REMS, that vaunted dream correlate, we have no marker beyond the dream report as to whether an individual is experiencing dreaming. Modern scanning systems such as fMRI, PET, and MEG, so useful in the study of REMS, turn out to be almost useless in studying dreaming.

Other biological correlates have been proposed for the various conscious states of mind. These proposed correlates range from gamma frequency to complexity, neural networks, quantum states, and neural cytoskeletons (17). These new theories have been constructed with little or no supporting evidence that these constructs are associated with dreaming. Dreaming, like many of the cognitive states, has no clear biological marker. Even worse, for the scientist dreaming occurs during sleep, a state defined by its lack of access to the waking world of both experience and experiment.

After 6000 years of religious, philosophical, and theoretical focus, it is only recently that experimentally testable empirical methodology, what is generally referred to as science, has been applied to the study of dreaming. It has not been easy to apply this approach. There are very few interested scientists, and an even more limited diversity of methodological techniques. There are powerful, if unproven, constructs of theoretical belief, and very little funding for scientific research likely to refute those theories. If those limitations were not enough, it is nature of the study of dreams to derail both logic and science. As David Foulkes points out, "... There is something about dreaming that has always seemed to move people prematurely to forsake the possibility of disciplined empirical analysis" (18). In perhaps no other area of modern scientific study have professionals been so willing to accept their presuppositions (e.g., REMS is dreaming) without the requirement of empirical evidence.

MELANCHOLIC SIMPLETONS

The situation was even worse in classical antiquity, when most everyone could set themselves up as a philosopher, a teacher, or an interpreter of dreams. No diplomas and no evidence were required. Theories supported the views of philosopher/scientists with large reputations. Such individuals would simply shout down, de-emphasize, or physically suppress contrary viewpoints. Aristotle was perhaps the first to bemoan the philosopher/scientists and their haphazard uses of empirical fact in support of their *a priori* theoretical views. The great theorist was never one to get out of the way when pontificating on his particular view of the world. In his time, in Classical Greece, dreams were the stuff of prophecy – evidence of contact with the gods. Aristotle was not Socrates. Valuing his health and safety, he argued that dreams were evidence for the existence and active involvement of the gods in human affairs, not as messages from gods outside, but as demonstrations of god's creation, the working mind, "Dreams are not sent by a god, nor do they exist for this purpose; however, they are beyond human control, for the nature [of the dreamer] is beyond human control, though not divine" (19). In searching for empirical evidence for this theory, Aristotle came up with a brilliant, poetic gem, in suggesting that simpletons, persons of limited intelligence, and those with melancholia were those most likely to have prophetic dreams. There was, of course, no evidence support his theory. But this construct presented a virulent line of attack that could be used on any melancholic simpletons still convinced that dream-based prophesy was a gift of the gods.

A SLICE OF THE BRAIN

In science, definitions applied to entities that are not available for direct observation are often confounded with the techniques used for measurement, since that technology can be objectively studied (20). It is far easier to study the electrophysiological maker of REMS than it is to study the cognitive state of dreaming. And does it really matter that REMS is not equivalent to dreaming? Today, nearly everyone believes that brain equals mind. Each month, another scanning study is published, illustrating a potential teleological brain site for consciousness, spirituality, love, or hope. Many cognitive scientists and many in the general public believe that a slice of the brain is responsible for thoughts, feelings, and dreams. Almost everyone is a monist, convinced that brain equals mind. Authors of biological consciousness theories have gone on to found entire fields of study, chair university departments, and obtain funding from the National Institutes of Health. They have been given grandiose awards by their namesake scientific societies. The famous "old men of dream sleep" have done very well. Many are now publishing fascinating memoirs. Some were perhaps driven less by science than by the need to support their theories. There have never been many grants given for the study of dreaming. Those that were given went to the research communities and departments supporting REMS neuroconsciousness theories. Researchers who chose to emphasize the evidence that dreaming occurred throughout sleep did not go on to found departments at major institutions or to receive prestigious rewards. Many dream scientists spent their careers quietly continuing to work in the hope of being published. The study of the actual dreaming state was virtually suspended as decades, money, and effort were invested in the study of the red herring of REMS – dreams' supposed biological marker. REMS is a fascinating biological state that is sometimes associated with dreaming. Dreaming is a complex mental state that reflects cognitive process, alters waking function, and potentially defines our species. Dreaming, when equated with REMS, is a story that has much missing.

AN ALL-TIME LOW

With the loss of the REMS equals dreaming correlate as a basis for neurobiological theories of dreaming, research into the dream state has gone through a nadir of activity. The percentage of scientific papers on the topic of dreaming is currently at an all-time low (see figure) (21). Recent published works on dreaming have documented steady incremental increases in the understanding of the state, rather than breakthrough insights into the meaning of existence and the origin of consciousness. This recent work reveals dreaming as an exceedingly complex state. There are electrophysiological, neurochemical, anatomical, and physiological systems of dreaming, with each of these systems affected by a wide variety of medical, psychological, sleep, and social variables. The dream report can be studied by addressing the reported cognitive processing and the observable characteristics of the dream (i.e., dream and nightmare recall frequency, methodologically controlled content, disease and medication effects, and dreaming effects such as emotional expression, creativity, learning, and memory). When dream science was centrally focused on REMS, there was less interest expressed in these other aspects of dreaming. Dreams were believed to originate from primitive aspects of the mind, and many of the cognitive and metacognitive aspects of dreaming were denigrated and ignored. Most researchers believed that they had a clear understanding of the biological basis for dreaming, so that today we understand less about dreaming than we

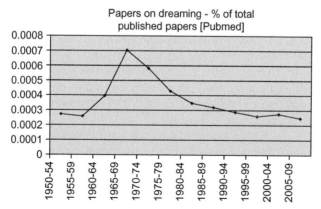

Papers on dreaming: percentage of total published papers (PubMed, 1950–2009).

understand about the other cognitive states. After 4000 years of interest and study, dreaming still eludes our attempts at explanation.

ALMOST UNSTUDIED

There is an upside. Dream researchers are faced with a broad horizon. Dreaming is universally experienced. It is potentially the characteristic state that defines our species. And scientifically dreaming is an almost unstudied and untapped resource. Modern tools of scientific study – real-time scanning systems using PET, fMRI, and SQUIB (Chapter 4) – are producing revolutionary changes in the fields of cognitive science. There is a wealth of research addressing REMS using PET, SQUIB, fMRI, brain-slice, and micropipette techniques, backed up by complex statistical methodology. Consequently, research has revealed much about the neuroanatomy, neurochemistry, and neuropathology of REMS, and very little about dreaming. As researcher and clinician Mark Solms has pointed out:

> The fact that not only humans but all mammals display the REM state made it possible for neuroscientists to go one step further: they could identify the brain mechanisms underlying the REM state (read: dream state) by means of animal experiments that were ethically unacceptable in humans. This is where the slippery slope began, for no matter how close the homologue may be between the REM state in humans and other mammals, we have no way of knowing whether the same applies to their dreams. The moment investigators switched from studying humans to other animals, the monitoring of their subjects' dreams (as such) was perforce abandoned. (*Solms and Turnbull, 2002, p. 184*) *(22)*

If further progress is to be made, dream scientists will by necessity need to return to the study of actual dreams. They will need to retrieve what are underutilized skills in this current era of technocracy: well-designed and reproducible experimental methodology, strict scientific method, explicit definitions, multiple and controlled collection modalities; clear, complete, and simplified statistical analysis; and empirical, logical study design. These approaches offer the potential for quantifiable and reproducible scientific insights into the dream state: a cognitive state that, addressed without presuppositions, is wide open for potential study.

My previous book, *The Limits of Dream* (2007), focused on how psychoanalytic and neuroscientific theories of brain functioning were affected by the loss of the REMS equals dreaming correlate. *The Limits of Dream* looked at how this impacted the study of dreaming, and the effects of this change in understanding on fields of study that had incorporated this belief system into their structure and theory. This book extends that attempt – looking at what we actually know of dreaming after discarding the dissociable correlate of REMS. The closure of the REMS equals dreaming era has left a wide opening. *Dream Science* presents the current scientific knowledge on the state of dreaming, while addressing the much larger and enticing canvas of what remains unknown.

PARADIGM SHIFT

For dream science, this is a paradigm shift, constituting what Thomas Kuhn described as a scientific revolution for the field (23). This shift affects all the associated fields of study that have incorporated a biological basis for dreams into their conceptual framework. These fields run a gamut that ranges from basic neuroscience to philosophy, and it is hard to find any that are not affected in some way by this structural change. Since this shift has only just occurred, the changes are not yet fully apparent; and this book will be one of examples, describing some of the fields where this shift in understanding readily applies. Viewing dreaming without REMS alters the landscape of consciousness. In exploring this story, let's begin at a time when the effects of dreaming on our species first became apparent.

Notes

1. Pagel, JF. My adaptation of the Harry Hunt comment: "It is as if dreams are trying to become genuine stories but are typically falling a bit short." Hunt, H. (1991) Dreams as literature/science: An essay. *Dreaming* 1: 235–242.
2. Freud, S. (1953) The interpretation of dreams, in *The Standard Editions of the Complete Psychological Works of Sigmund Freud*, Vols. IV and V (1907), ed. J. Strachey. London: Hogarth Press.
3. Freud, S. (1917) Psychoanalysis and psychiatry general theory of the neuroses, in *Introductory Lectures on Psychoanalysis* (1966), trans. and ed. J. Strachey. New York: W. W. Norton, p. 255.
4. Freud S. *New Introductory Lectures on Psychoanalysis*. Harmondsworth: Penguin; 1933/1973. p. 83.
5. Pagel JF. *The Limits of Dream – A Scientific Exploration of the Mind/Brain Interface*. Oxford: Academic Press/Elsevier; 2008.
6. Berger H. On the electroencephalogram of man: Third report. *Electroencephalography and Clinical Neurophysiology Supplement*. 1931/1969;28:95–132.
7. Dement W, Vaughan C. *The Promise of Sleep*. New York: Dell; 1999: pp. 35–36. William Dement, Professor of Psychiatry and Behavioral Science at Stanford, basically founded the field of Sleep Medicine. In 2013, at 86 years of age, he was still attending the national meeting of the American Academy of Sleep Medicine. He's a remarkable and intelligent man.
8. Pinker S. *How the Mind Works*. New York: W. W. Norton Co.; 1997. p. 77.
9. McCarley R, Hobson J. Neuronal excitability modulation over the sleep cycle: a structural and mathematical model. *Science*. 1975;189:58–60. Robert McCarley, a professor of Psychiatry at Harvard, has never tightly defended this model. He recently asked the author of the current text (no fan of REMS equals dreaming) to write the introductory chapter to his newest book [*Rapid Eye Movement Sleep – Regulation and Function*

(2011), ed. B. N. Mallick, S. R. Pandi-Perumal, R. W. McCarley and A. R. Morrison. Cambridge: Cambridge University Press].

10. Pace-Schott EF, Solms M, Blagrove M, Harnad S, eds. *Sleep and Dreaming: Scientific Advances and Reconsiderations*. Cambridge: Cambridge University Press; 2003.

11. Hobson J. *Dream Life: An Experimental Memoir*. Cambridge, MA: MIT Press; 2011. J. Allen Hobson, Emeritus Professor of Psychiatry, Harvard Medical School, built the Department of Psychiatry at Harvard into a defensive bastion for activation-synthesis and REMS equals dreaming.

12. Solms M. Dreaming and REM sleep are controlled by different brain mechanisms. In: Pace-Schott EF, Solms M, Blagrove M, Harnad S, eds. *Sleep and Dreaming: Scientific Advances and Reconsiderations*. Cambridge: Cambridge University Press; 2003:51–58. Mark Solms, a lecturer in neurosurgery at the Royal London School of Medicine, was declared winner of the famous debate with Hobson on REMS and dreaming (based on a show of hands) at the Towards a Science of Consciousness meeting in Tucson (2006). He now spends the majority of his time at the University of Cape Town, South Africa.

13. Foulkes D. *Dreaming: A Cognitive–Psychological Analysis*. Hillsdale, NJ: Lawrence Erlbaum Associates; 1985. David Foulkes did this very early and excellent work demonstrating that dreaming occurs during all sleep stages. He rarely, if ever, attends meetings in the field.

14. Pagel JF. REMS and dreaming – historical perspectives. In: Mallick BN, Pandi-Perumal SR, McCarley RW, Morrison AR, eds. *Rapid Eye Movement Sleep – Regulation and Function*. Cambridge: Cambridge University Press; 2011:1–14.

15. Nielsen T. A review of mentation in REM and NREM sleep: "Covert" REM sleep as a possible reconciliation of two opposing models. In: Pace-Schott EF, Solms M, Blagrove M, Harnad S, eds. *Sleep and Dreaming: Scientific Advances and Reconsiderations*. Cambridge: Cambridge University Press; 2003:59–74. Tore Nielsen, director of the Sleep Research Center at Hospital du Sacre-Coeur de Montreal, is the latest apologist for REMS equals dreaming, the recipient of multiple Canadian grants for the study of dreams, and the preceptor of the majority of researchers trained to study dreaming in North America.

16. Pace-Schott E. REM sleep and dreaming. In: Mallick BN, Pandi-Perumal SR, McCarley RW, Morrison AR, eds. *Rapid Eye Movement Sleep – Regulation and Function*. Cambridge: Cambridge University Press; 2011:8–20. With Hobson's retirement Edward Pace-Schott holds down the Harvard fort.

17. Searle JR. *The Rediscovery of the Mind*. Cambridge, MA: MIT Press; 1998.

18. Foulkes D. Functions of dreaming. In: Moffitt A, Kramer M, Hoffmann R, eds. *The Functions of Dreaming* Data constraints on theorizing about dream function. Albany, NY: SUNY Press; 1993. pp. 11–20.

19. Aristotle On divination in sleep. In: Van Der Eijk P, ed. *Medicine and Philosophy in Classical Antiquity: Doctors and Philosophers on Nature, Soul, Health and Disease*. New York: Cambridge University Press; 2005:189.

20. Garner W, Hake H, Eriksen C. Operationism and the concept of perception. *Psychological Review*. 1956;63:149–159.

21. Pagel, JF. Preface, in *Dreaming and Nightmares*, ed. JF Pagel. *Sleep Medicine Clinics* 5 (2): xi–xiii. Philadelphia, PA: Saunders/Elsevier; 2010.

22. Solms M, Turnbull O. *The Brain and the Inner World: An Introduction to the Neuroscience of Subjective Experience*. New York: Other Press; 2002. p. 184.

23. Kuhn T. *The Structure of Scientific Revolutions*. 2nd ed. Chicago, IL: University of Chicago Press; 1962.

1

Archaeology, Anthropology, and Dreaming

None of us could paint like that … We have learned nothing in 12,000 years. (Picasso) (1)

We are a dreaming species. Coming out of sleep, we experience the finding of our way into wakefulness, to arrive at a place where almost all of us remember the thoughts, images, and emotions that occurred during our sleep, labeling them as what we call dreams. We are not alone in this experience. The human species, as far as we can know, has always dreamed, at least since the dawn of recorded history and probably far before. The Paleolithic cave paintings that flickered in the first artificial firelight have much in common with dreams. Like dreams, the unreal magical creatures on the walls of the caves are images seen through the lens of a belief system, incorporating memories and emotions. And like dreams they can affect others.

1

The point at which we began to dream may be the point at which we became recognizably human. A dream is a waking realization of internal and non-perceptual mental processing. In order to dream, we must have acquired the capacity to view ourselves as independent of one another and the world around us. In order to dream, we must have developed the ability to consider, in a form of self-reflection, the inner working of our brains and bodies.

Dreaming marks the presence of consciousness awareness, not just of the world around us but also of ourselves as independent functioning creatures. It is likely that all *Homo sapiens* have woken from the strange death-like state of sleep to the same jumble of images, thoughts, and emotions. These dreams form a mental reflection of our waking life experience. Using the tools of language and grammar, we take the continuity of waking experience from the day before, combining the mental activity occurring during sleep, and form it into stories. And, oh, what stories!

MOVING IMAGES - CAVE ART

Before the advent of written language, in the caves of southern France and Spain, we drew mystical creatures: pregnant mares fat with possibilities, now extinct ibexes nodding exquisite swept-back horns, dancing cave bears, mastodons, and charging rhinoceroses. As Levi-Strauss has pointed out, our only access to the Paleolithic mind is through close analysis of the products of those minds (2). Beyond unadorned stone tools, a few pierced beads, stringed shells, and a red ochre scratch design found in South Africa, there are few other decorated human artifacts that pre-date the paintings of the Chauvet Cave in France from 32,000 years ago (3). These Paleolithic images were created by humans able to create representative two-dimensional images independent of their creators and the world around them.

These paintings mark a period of major and accelerated changes in the development of the human species. Between 30,000 and 50,000 years ago, there is the first evidence that our human ancestors developed refined and decorated tools for use beyond the utilitarian. We find the first evidence for body adornment and for the elaborate burials for certain individuals. In this area of southern Europe we find the cave paintings, the first images (4). This is also the period in which the specific strain of the *Homo* species that was to become our ancestor differentiated from the other proto-human species such as *Homo neanderthalensis*. Genetic evidence indicates that the Neanderthal contributed at least minimally to the modern gene pool even as remnant elements were in the process of becoming extinct (5). It is possible that the other proto-human species also had representational art, perhaps created on external rock outcroppings that were destroyed by rain and wind, or on perishable materials long lost to the historic record. There is little question that the art being drawn on the walls of hidden caves had a major role in preserving what still exists today. The available evidence does indicate that the paintings were created by our ancestors (6). As far as is currently known, none of the other proto-humans, including the Neanderthals, produced depictions (7).

It is not fully clear why these paintings were created in only this area (8). By this point in time archaeological evidence indicates that various *Homo* species had disseminated throughout Africa, Asia, Europe, and Australia (9). There is also the question of function: why were these images created? The arguments having to do with the value or function of artistic creations persist to this day. In a world in which starvation based on declining resources, illness, predation, and internecine conflict is a real possibility, the effort involved in learning, training, and creating art is arguably misapplied. This perspective further emphasizes the importance of addressing the potential function of the cave paintings. From the perspective of dream science, it is hard to overemphasize the importance of the cave paintings.

- The cave paintings are the first clear archaeological evidence that our species had obtained the capability of reflexive consciousness.
- Reflexive consciousness is required for and associated with the capacity to dream.

Most animals have a form of primary consciousness, an awareness of the outside world. Several species, including elephants, bottlenose dolphins, and apes, demonstrate an ability for self-awareness that may rival that of humans (10). Proto-humans clearly had a primary consciousness of the external world, and were likely to have been self-aware, with the capacity to view themselves as independent of one another and the world around them. Without self-awareness, they could not have conceived of themselves as independent organisms, somehow existing apart from the external world. Modern humans extend that awareness into what has been called "reflexive consciousness" – the ability to self-consider the working of our brains and bodies. Reflexive consciousness is required if we are to organize our thoughts independently of others, and includes the recognition that the thinking subject has his or her own acts, and the existence of a socially based self-hood that affects others on a timeline that includes the personal, the past, and the future (11). These capabilities require an awareness of our own thoughts and an understanding of the behaviors of others. It has been suggested that this is the capability that differentiated our ancestors from other proto-modern humans (12). The paintings potentially mark the moment when people began to conceive of themselves as different from animals – the moment when they became human (13). Some authors extend this perspective, suggesting that the marker for the development of reflexive consciousness may have been the development of dreaming (14).

The cave paintings are also evidence that their creators had acquired the capacity to create two-dimensional representations of three-dimensional

TABLE 1.1   The Operative Cascade of Visual Imagery

Picture	Develop your pattern and configuration map of the image
Find	Use attention to shift image properties and coordinate patterns
Put	Focus on the description and relationship of a part to the whole image
Image	Establish object names, size, location, orientation, and level of detail
Resolution and regeneration	Delineate the comparative detail of this image
Look-for	Integrate relevant memories
Scan, zoom, pan, and rotate	Your presence, your operative attention in the image
Answer-if	Do properties associated with the image answer an already developed cognitive search parameter?

Derived from Kosslyn (1994) (16).

objects. This ability, like dreaming, also requires the capacity for reflexive consciousness. Visually, dream images are representationally two-dimensional (15). It is interesting to suggest that the ability to create two-dimensional representations is related to the capacity to experience dreaming. Of course, the cave paintings may have had nothing to do with dreams. The Paleolithic mind cannot be read, and there is little empirical evidence for this supposition. The evidence we do have can be summarized as follows.

- Visually, dream imagery utilizes a cognitive paradigm that includes the transformation of images from three- to two-dimensional status, incorporating aspects of point of view, perspective, and placement of imagery (Table 1.1).
- An identifiable primary function of dreaming is in the creative process, and most artists utilize dream input in their work.

To further explore this theory would require finding a group of individuals who have difficulty creating two-dimensional representations of three-dimensional objects. This theory might prove correct if it could be demonstrated that these individuals also have a change in their ability to dream or exemplify differences in the style and content of their dreams. This possibility seems worthy of further explanation when one considers that dreams are frequently used for visual art, creativity, and science.

From the dawn of recorded history, extending into the science, creativity, and art of this era, part of being human has been to be fascinated by dreams. Historic hunter–gathering societies have also demonstrated the same fascination, using their dreams in concert with painted imagery in ceremonial and religious practice. However, it is clear that such an ethnographic approach can lead to flawed perspectives of ancient behaviors. Since modern primitive tribes sometimes use the images of prey animals in ritual preparation for hunting, the paintings were initially viewed from the perspective of functioning in the process of hunting magic or "wish fulfillment." The animal images were viewed as totems and pictures of animals that the hunter might want to multiply and successfully kill (17). The wish-fulfillment component of this perspective also fitted with the Freudian psychoanalytic theories of the time. Even so, the hunting magic theory is likely to be incorrect. There is little correlation between depicted animals and food remains at the site, and this once predominant theory seems to explain little about the paintings or their function in the Paleolithic world. After a history of such potentially misapplied ethnology in which the actions of modern "primitive" groups were attributed to ancestors and other proto-humans, anthropologists now argue that ethnographic comparisons can be considered legitimate only when used to first formulate new ideas, or in the special situation when a characteristic is universal and found among all known historic and modern "primitive" societies (18). Dreaming meets both of these criteria.

The cave paintings, hidden deep in caves, were visited recurrently by the groups that made them. Sites of particular significance were visited repeatedly over tens of thousands of years (19). A majority of the pictures are located in absolute darkness and require imported light for viewing. Most of the pictures are of animals. There are only rare human figures and almost no landscapes. Some of the animals were human prey, while others are images of those that preyed on humans. These animals were obviously of importance, part of the external world and belief constructs of Paleolithic hunter–gatherers. The artists incorporated perspective, color, and point of view. Some of the animals are depicted in multiple positions and appear to move. Many of the paintings utilize aspects of the cave site, incorporating natural characteristics of the cave wall to form and bring out three-dimensional aspects of the figures, and utilize line-of-sight tunnel perspectives to frame important paintings. There is a sense of action: oxen straining with bulging muscles, lions on the hunt with gleaming eyes, the drama of galloping horses. Many are panels of multiple figures that present a storyline for the viewer showing the interaction of animals, intentional superimpositions, and sometimes their involvement with humans. These images are artistic, not realistic. They are stylized representations. The patterns of lines and shapes used to portray particular species were used consistently at widely separated sites, and presented in the same fashion over tens of thousands of years. Over many years using the same consistent and rarely varied style and techniques, many different artists created similar paintings. This indicates that the skills of painting must have been taught (20). While the animals depicted changed over time as some species flourished and others became extinct, the artists continued to use the same set of tools and pigments, the same techniques, and the same perspectives and styles of presentation (21).

The artistic skill levels demonstrated in these drawings are remarkable (see Picasso's opening quote). Artists including Picasso, the surrealists, and the post-impressionists have been confounded and impressed by their quality of technique. Those fortunate enough to see them cannot help but be moved and impressed; the power that they exert on the viewer today can be no less than the power they had to affect the Paleolithic viewer who journeyed into these deep sites, carrying ephemeral lights into the absolute darkness, confronting imagined images from far beyond their waking reality.

The sources for these images must have extended beyond waking experience. Although particular animals can be recognized, representations are stylized and non-realistic. They are dream-like in their form. The early history of modern societies and religions resonates with stories of dreams and religious trances that contributed to the stories and founding myths of those religions. Such an exploration of states of shifting consciousness is a characteristic of hunter–gathering societies that have been able to persist into the modern era (22). This is particularly true of the use of dreams, an approach incorporated into all human societies and all human religions (23). Some societies have also utilized psychoactive substances in order to obtain an alterative view of perceived reality, and in almost all modern societies, despite potential adverse consequences, psychoactive drugs are incorporated into the cultural milieu (24). But it is dreams that are almost universally available. Even today, many of us use our dreams in our decision making, our relationships, our work, and our play (25). Dreaming is a valued part of human experience in virtually every cultural community populating this planet. Almost everyone dreams, and those dreams provide a potential source of visionary insight and creative inspiration (26).

While today, we may only rarely dream of cave bears, the content of our dreams has similarities with the images drawn on the walls of French caves. As in dreams today, dream content, more than anything else, reflects our waking experience. What has clearly changed over 35,000 years is the content of waking experience. While our world has experienced vast changes, much about our dreaming has stayed the same. On the flickering wall of the cave are drawings of the horses, rhino, deer, and ibexes, the creatures of that waking life – the creatures of those dreams. Our dreams are filled with the modern waking experiences of gaming, films, and television. The dreams of the ancients proto-humans were filled with the now extinct animals that defined the experience of living in that world. The experience of dreaming was probably much like the dreaming that you and I do each night and sometimes remember. Each dreamer, upon waking, is confronted with a set of thoughts and images that are seemingly meaningful, and seemingly important, integrating the personal history and experience of very different worlds.

Our belief systems affect our dreams, and it is likely that the animals dreamed and drawn by the ancients were viewed far differently from how we view animals today. In the 1950s, with the advent of general television viewing, most people reported that they dreamed in black and white (27). Such a dream report today is considered unusual and meets criteria for being significantly bizarre (28). Today, all dreams are in color. Before black and white television, it was probably quite rare for anyone to dream in black and white. And no one thought to ask whether dreams included color. Why would they not? The animals of the cave paintings are impossibly ancient, giving insight into a Paleolithic mind so different from our own that in many ways these were the thoughts of a different species. But that species dreamed, those dreams reflected their world, and those dreams were used to change that world until eventually, after tens of thousands of years, that world became the world in which we dream today.

MIMEOGRAPHS ON FRESH CLAY

Dreams have changed the world. Dreams have inspired conquerors, prophets, and the writers of our ancient holy books. Dreams that were, at their origin, someone's own personal dream, no different from the dreams that come to each of us, have led to new religions, philosophies, and scientific discoveries. Each one of us who dreams first connects with the concept of dreaming by integrating our personal history and experiences of dreaming. We look to see whether our dreams are similar to the dreams of others – the dream that led to marriage, to choice of career, to a work of art, or to a new understanding of a relationship with our particular god. Once we start to ask about dreaming, we become more likely to remember our own dreams. As we ask, we explore and read, we begin to see how our dreams are the same yet different from one another and from those of others. Eventually, we can explore what they might actually mean.

Once humans began to write, among the first things that they did was to record their dreams. Some 4000 years ago in Lagash, a city of ancient Sumeria, round fired-clay codas were created, designed to be rolled like an ancient mimeograph on fresh clay to create multiple imprints. Among lists documenting agricultural production and supplies, and a listing of ancient kings, archaeologists found imprints of a dream that had come to their King Gudea. That coda reports his dream story:

> In the dream, the first man – like the heaven was his surpassing size, like the earth was his surpassing size, according to his horn-crowned head he was a god, according to his wings he was the bird of the Weather god. According to his lower parts he was the Storm-flood, lions were lying to his right and left – commanded me to build his house, but I do not know what he had in mind. Daylight rose for me on the horizon. The first woman ... held in her hands a tablet of heavenly stars she put on her knees consulting it. The second man ... holding a tablet of lapis lazuli in his hand, set down the plan of the temple. Before me stood a pure carrying pad, a pure brick mold was lined up, a brick determined as to its nature, was placed in the mold for me, in a conduit standing before me was a slosher, a bird-man, keeping clear

water flowing, a male donkey at the right hand of my lord kept pawing the ground for me. *(Oppenheim, 1956, pp. 245–246) (29)*

A priestess's interpretation of this dream is also recorded. The dream was interpreted as giving the instructions of how to position the king's new temple according to the cardinal points, and specifying on which auspicious day the work should begin (30). There are codas with greater archaeological import, relics from older sites, in sometimes unreadable scripts, and even a long comparably ancient history of painted Egyptian tombs that contain minimal references to dreams. But alongside the lists, the gods, the produce orders, the food and processions destined for the afterlife, tax records, and the hierarchical lineage of ancient kings, among our first documented attempts at a written language, there is perhaps the first attempt at the written communication of story – a story related as a dream.

It is in the world's religious and spiritual traditions that we can find the oldest documented dreams. The best records of historically ancient dreams and the dreaming state are those incorporated into ancient religious texts (31). These first recorded dreams reveal tensions, conflicts arising from our attempt to understand our place in the world. Men, women, kings, paupers, lepers, priests, the crazy and the sane alike, all had dreams. Whether these dreams came from humans or gods, dreams came during the night from the dark realm that we call sleep – a state of consciousness that from the outside looked so much like death; a state that differed from waking reality, filled with content that was sometimes unreal and exceedingly bizarre. These dreams inspired people, leading to creatures that were drawn, clay that was shaped, and stories that were told. Were these dreams the interpersonal experience of this creature dwelling for a limited time on a tumultuous and often difficult world, or were these dreams of something seemingly larger – something from outside, something from what we came to call our gods? If we chase the story back, to the limits of written records and beyond, this is how the study of dreams began. What was true, what was false, what was ours, and what came from others? Those questions led down intellectual pathways, leading first to religion, then to philosophy, and eventually to science.

This long history of the incorporation of dreaming into the world's religious and intellectual texts suggests that it may actually be important for us to ask what is known about the dream state. Are we any closer today than we were 4000 or 32,000 years ago to an understanding of what a dream is? Do we have any idea of what a dream might mean?

NIGHT TERRORS

When I was a child, I would wake screaming from a recurrent vision that would come to me during sleep: darkness rumbling gray – wet, rock wheels set moving together in circuit; green moss covering wheels that slowly roll together in a room deep underground. No matter how hard I tried to hold a tight focus, I would always falter and in that flicker of inattention, the wheels would begin to totter and lose their synchrony, their easy rolling. The rock wheels would crash into one another in eerie silence, smashing and shattering chamber walls, breaking out of darkness, sending me screaming out of sleep. I would wake both frightened and nauseated, myself at eight years, lying there in bed covered in a cold sweat, amazed. Fifty years later, much of my childhood is lost, but not that dream. Remembering it still makes me feel queasy.

Such is a dream, and without any outside input, if pressed, I can understand the allusions, even some of the meanings, associated with the content of my dream. A Freudian might see this as a memory of the trauma of birth. Jung's soul circles are here deep underground moving in unconscious synchronicity. Potentially, these are profound insights with personal resonance. But interpretations address only a portion of the content of this dream experience. In the dream, there are colors, tactile associations, and associations that later in my life will induce the experience of *déjà vu* – the illusion of having previously experienced a dark and underground location similar to that in my dream when I was actually encountering a site for the first time. This dream induced symptoms of physical discomfort, and the confusion of arousal into waking. This dream comments on my relationships, my developing sexuality, and the way my mind operates. The potential personal associations for this dream, the available metaphors, are semiotic, leading to infinite possibilities for interpretation. Could such a dream have a function? Could it have meaning? Or are dreams meaningless thoughts that occur during our unfocused unaware states of sleeping? How does that dream apply to me, the person having the dream, the person to whom it matters?

We have the option of looking outside our dreams to records documenting the evidence-based understanding of dreaming – what we will call "dream science." Dreams may actually be important, and perhaps surprisingly without the support of the dissociable biological marker of rapid eye movement sleep (REMS), dreams and dream science can stand alone. Particularly in those fields that had based a portion of their framework on the "REMS equals dreaming" correlate, this reconsideration of dream science is having considerable impact on both theory and application.

This debate extends to the question of dreaming in the Paleolithic era. Since most animals and all mammals have REMS, it is unreasonable to propose that any species of great ape did not have REMS. If REMS is dreaming, it is difficult to postulate that any proto-human, ancestor or not, would not have experienced dreaming. If REMS is dreaming, any psychoanthropological differentiation between other proto-humans and ancestor *Homo sapiens* based on the presence and/or absence of dreaming is doubtful based on the biological evidence. If dreaming is rather a cognitive process that entails the recall of personal cognitive processing occurring during sleep, the dream state could potentially mark the revolutionary differentiation of our species from others. The ability to dream could reflect the development of reflexive consciousness and the development of two-dimensional paintings, elaborate burials, and complex language by our *Homo sapiens* ancestors (32).

Archaeologically, the cave paintings mark the point at which an accelerated process of change occurred for our species relative to others. In reviewing the archaeological evidence on Neanderthal cognitive processing, Wynn and Coolidge have come to the conclusion that the only difference between the Neanderthal approach to technology and ours is in the area of innovation. The Neanderthals almost never innovated. In their 160,000 years of known archaeological history, there is only one clear example of innovation: the invention of hafting – the ability to tie a stone point to a spear. Their inability to innovate may have negatively affected their ability to adapt and compete with our Paleolithic ancestors (33). Our ancestors were different in this way, and today rapid technological change still characterizes our species. This species characteristic of innovation and creativity is at least in part secondary to our cognitive capacity for dreaming.

Dreaming, an altered state from waking consciousness, provides an alternative perspective of reality, requiring us to organize thought in a way that differentiates the apparent world of dreaming from the perceptual reality of the outside world. Dreaming provides information that can be used in developing alterative approaches to functioning in the external world. Dreams can be integrated into our creative process, allowing for the possibility of creative products such as cave paintings. Dreaming supports individual and species-based creativity.

Creativity has class 1 survival value (Table 1.2) and may be among the most important cognitive capacities characterizing our species. Dreaming, in its involvement in the creative process, would be considered to have at least class three survival value (35).

There is excellent evidence that a primary function for dreaming is creativity (36). Anecdotes abound for the association of dreaming with scientific discovery, from Descartes' discovery of the scientific method in a dream to Kekulé's discovery of the benzene ring. Creative process and dreaming are both self-expressive experiences that diverge from the stepwise, perceptual consciousness in which we live our daily lives. Each person has a unique concept of which process is dreaming and which process is creative. Dreams are often interpreted and creativity is viewed best when addressed as global, gestalt experiences. Intense dreaming and creativity are associated with similar personality types (openness and thin borders) and psychiatric dysfunctions (schizotypal personality and bipolar disorder) (37). Both have been the focus of psychoanalytic thought.

TABLE 1.2 Classification of Functional Significance for a Survival Characteristic

Class	Description
1	The characteristic has a survival characteristic at that point in development
2	The characteristic does not have a survival characteristic at that point in development, but it is necessary to build another characteristic that eventually will have high survival value
3	The characteristic does not have a survival characteristic at that point in development, but it will eventually be combined with other characteristics, the combination eventually having high survival value
4	The characteristic itself does not have a survival characteristic but it is the outcome of that characteristic that falls in one of the first three categories
5	The characteristic does not have high survival value

Adapted from Fishbein (1976) (34).

Both are among the most individual of experiences. Each allows the opportunity for incorporating alternative and unexpected associations into waking thought and functioning.

THE MECHANICAL "CREATIVENESS" OF A LIGHTNING STORM

Creativity is difficult to define and difficult to test. The term, derived from the Latin base *creatus* ("to make or produce" or literally "to grow"), was first used in literature by Herman Melville in *Moby-Dick* (1851) to describe a lightning storm (38). In the nineteenth and early twentieth centuries, the study of creativity was called the study of "genius" and "imagination." Webster's dictionary (1998) defines creativity as artistic or intellectual inventiveness (39), while Random House (1989) describes it as the ability to transcend traditional ideas, rules, patterns, relationships or the like, and to create meaningful new ideas, forms, methods, interpretations, etc. (40). The internal conditions for constructive creativity are considered to be:

- openness to experience including extensionality, which is the opposite of psychological defensiveness, rigidity, and rigid boundaries
- an internal locus of evaluation in which the value of a creative person's product is established by oneself
- an ability to toy with elements and concepts, to juggle elements into impossible juxtapositions (41).

Creative problem solving is often based on an ability to go off in different directions when faced with a problem (42). From a "self-expressive" perspective, creativity has been defined as "the ability to think in uncharted waters without influence from conventions set up by past practices" (43) and "the process of change, of development, of evolution, in the organization of subjective life" (44). Making variations on a theme can be considered the crux of creative thought. Creativity is an externalized exploratory approach that we can utilize in philosophy, art, and science. Creative thought is an investigation of the structures and tensions existing on the boundary of body and mind. In order to function optimally, creativity needs an access, a window, between mind and brain. The dream is that access, a tool for use in the creative journey of exploration.

Creative process has traditionally been considered to include the production of new and original products. But pastimes such as golf, hunting, reading, relaxing, and even television watching are described by some as the personally creative parts of a creative life (Table 1.3) (45).

TABLE 1.3 Self-Reported Types of "Primary Creative Outlet" ($N=424$)

Experiential Creative Outlets ($N=187$)		Traditional Creative Outlets ($N=204$)
Archery	Housework	Art
Basketball	Hunting	Calligraphy
Bowling	Lobbying	Coloring
Camping	Motorcycling	Computer
Cars	Puzzles	Crafts
Child	Ranching	Crochet
Church	Reading	Dancing
Construction	Religion	Doodling
Electronics	Remodeling	Drawing
Family	Resting	Music
Fishing	Running	Painting
Games	Shooting	Photography
Gardening	Soccer	Piano
Genealogy	Sports	Quilting
Golf	Tanning	Sewing
Grandparent	Tinkering	Theater
Hiking	Watching TV	Woodworking
History	Weightlifting	Writing
Horses	Yard work	

ILLUMINATION

Many modern artists and writers attribute at least a portion of their creative inspiration to their dreaming. Dreams have historically been included as part of the paradigm for the process of creative problem solving. This process described by Wallas in 1926 has four parts: preparation, incubation, illumination, and verification (Table 1.4) (47).

The dream is the illumination. This approach has a long history of use in both ancient and modern societies. The individual will go to sleep thinking of a problem with his or her work, and will often wake having discovered an alternative approach. Today, many writers and artists report using forms of dream incubation in their creative process (48). Colin McGinn suggests that such experiences of the imagination are required for the creative process. He suggests that there is a spectrum that denotes the imaginative experience that starts with a memory image that is expanded into the process of imaginative sensing to produce a productive image that formed as a daydream/dream and eventually results in meaning and creativity (Table 1.5). He has proposed that this is the dynamic that we use to transition from the sensory experience to cognitive imagination in order to form a story.

Successful film makers, actors, and screenwriters use their dreams at a far higher frequency than the general population. They are more

TABLE 1.4   Steps for Utilizing Dream Incubation in Creativity

Step 1	Choose the right night – when you are not overtired or inebriated and you have ten or twenty minutes to prepare before sleep and ten minutes, at least, for recording in your journal the next morning
Step 2	Keep a journal – write down your day notes in the period before sleep; this will relax you, clear your mind, and orient you towards your journal
Step 3	Incubation, discussion – address with your conscious mind the problem you need to resolve, reviewing causes, alternative solutions, your feelings, and possible secondary gain involved in resolving the problem
Step 4	Incubation phase – write down a one-line question or request in your journal that expresses as clearly as possible your desire to understand the dynamics of your predicament (make this statement as simple as possible)
Step 5	Focus – put your journal beside your bed, turn off the light, and close your eyes. Focus your attention and concentrate on your question, repeating the phrase in your mind until you fall asleep
Step 6	Sleep – usually the easiest step
Step 7	Record – all dream memories should be written down in your journal as soon as you wake, either during the night or first thing in the morning

From Delaney (1979) (46).

TABLE 1.5   The Process of Imaginative Experience

Memory image

→ Imaginative sensing

→ Productive image

→ Daydream/dream

→ Possibility and negation

→ Meaning

→ Creativity

From McGinn (2004) (49).

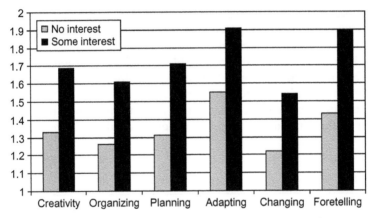

FIGURE 1.1   Dream use frequency varying from 1 = never to 5 = always, compared between no creative interest (*N*=58) and the group reporting some creative interest (*N*=424).

likely to recall both dreams and nightmares (50). It is not only successful artists, writers, and scientists who use dreaming in their creative process. When asked to identify their creative process, our sleep laboratory patients described physical activity (23 percent), crafts (15 percent), and music (15 percent) as their creative process. Nineteen percent reported having no creative interest. We found that if an individual had a creative process, whether it was gardening, sports, or listening to music, that individual had a significantly higher level of dream use compared to the group with no creative interest. And dreams affected many aspects of waking behavior that extended beyond the use of dreams in creativity (51) (Figure 1.1).

NON-DREAMING NEANDERTHALS

Another line of evidence supporting the association between dreaming and creativity comes from a study of non-dreamers – individuals who never dream. In our sleep laboratory, 6.5 percent of patients (35 of the 598 questioned) stated that they did not dream. In follow-up telephone interviews with the twenty-seven of these non-dreamers who could be contacted, five said that they had been mistaken and that they really did dream, six had dreamed as children but not as adults, and six had dreamed into adulthood but had lost that capability. Eight had not been dreaming when they had completed the questionnaire but had begun dreaming after treatment of their sleep disorder (in these cases it was the treatment of sleep apnea or depression). Only two continued to assert that they had never had a dream. This indicated that true reported non-dreaming was very rare in the sleep laboratory population, occurring in only 0.38 percent – one of every 262 patients. But it apparently existed.

In the sleep laboratory the classic way to determine whether someone is dreaming is to wake the individual up from different stages of sleep during the night and ask whether he or she has experienced thoughts, imagery, or emotions. This approach was used for these patients who reported by questionnaire and interview that they did not dream. Over a five-year period, sixteen such individuals were recruited for the study. In thirty-six awakenings, from both REMS and non-REMS, not a single patient reported a dream. As an experimental control, the same approach was used in a group who very rarely dreamed. The average time since their last reported dream was seventeen years. In three awakenings from three different individuals, a dream was reported. For two individuals that report was the sensation that a dream had been occurring. The other patient reported both content and storyline: he had been walking naked along the Left Bank in Paris. This individual insisted that although this realization had occurred during his sleep, this was definitely not a dream (Table 1.6).

The true non-dreamers in this study had no obvious memory impairment or difficulty functioning in society. Most had productive jobs; one was a mathematics professor. The one documented behavioral difference for this group of non-dreamers was their lack of interest and involvement in the creative process (52).

Based on the theory that the capability of dreaming is what differentiates our species from other proto-humans, these non-dreaming individuals are functioning as if they were proto-humans (Neanderthals?) in modern society. Counterintuitively, it is reported that the historic Buddha, after achieving enlightenment, never again dreamed (53). Even today, among some Zen Buddhist monks, the loss of dreaming is sometimes

TABLE 1.6　Non-Dreamer Study

	Individuals Who Report Never Experiencing Dreaming (N = 16)	Rare Dreamers (Mean Latency to Last Dream = 17 Years) (N = 12)
Total awakenings	36	32
Stage 1 and 2	18	22
– Dream reports	0	2 (9%)
Stage REMS	15	10
– Dream reports	0	1 (10%)
Stage 3 and 4	3	–
Total dreams reported	0 (0%)	3 (10.6%)

considered a marker for achieving enlightenment (54). The southern European cave paintings are located at the primary site and time of transition from a world occupied by a variety of proto-humans to a single almost modern hybrid *Homo* species. They may well be the creative reflection of a genetic, social, and cognitive hybridization between species that experienced the world in different yet complementary ways.

It may be that our primary ancestors dreamed while the Neanderthals did not. A cognitive approach including reflexive consciousness and dreaming would have been useful for the creation of decorated tools, elaborate burials, and especially their artistically stylized, beautiful, and ritually tied cave art. The archaeological findings from Neanderthal sites indicate that they had a non-innovative process of cognitive processing and a utilitarian approach to creating tools, burials, and art, quite different from the approach utilized by our ancestors. The remarkable cave art that flourished during the interface of contact emphasizes the cognitive differences between the species.

Dreaming versus non-dreaming is one way of expressing the differences in consciousness that existed between the *Homo* species. Wynn and Coolidge suggest that the difference in innovation between the species occurred owing to differences in working memory capacity that restricted the Neanderthals' ability to use their dreams in creativity. However, they have tied themselves to the belief that most dreaming occurs during REMS, particularly that used in the creative process. They also present an alternative proposal that while Neanderthals were capable of creatively using dreams, they did not do so. Reflective of their waking world, their dreams were so emotionally negative and full of aggressive content that they could not be used (55). This proposal suggests that the entire species had a variant of post-traumatic stress disorder, and avoided using their dreams and nightmares. This alternative proposal has significant limitations: many creative dreams are nightmares and many are associated with trauma (56). In later chapters we will address the creative nightmare.

The simpler proposal is that, as in all mammals, both species had REMS, and Neanderthals, like modern non-dreamers, had the anatomical and electrophysiological capacity to dream. The major differences between the species are in the type and form of cognitive processing. Cognitive archaeology indicates that the Neanderthals had a limited capacity for creativity and innovation. If they did dream, they did not use that capacity as we use it today – in the creative process. Perhaps, as Wynn and Coolidge suggest, this is due to a difference in available "working memory" that precluded their use of dreams. Working memory is another way of referring to the representational visual memory systems utilized in waking perception and imagery, as well as in dreaming. The evidence for a difference in working memory between species is based on the lack of evidence for the depictive use of symbols by Neanderthals – their lack of art. The representational memory systems are an extremely important component of both dream-like waking imagery and dreaming (see Chapter 9). If Neanderthals did not develop the cognitive ability to utilize working (representational) memory systems, it is likely that they could not dream, at least not as our species dreams today.

It is likely that the Neanderthal did not use what they called dreams in the creative process. But it is doubtful that this difference is based on differences between the species in the biological system of working memory. Monkeys and great apes utilize representational (working) memory systems in cognitively processing waking perception (57). It is likely that the Neanderthal, among the greatest of the great apes, also had this cognitive processing system. Dreaming includes multiple forms of consciousness and involves multiple memory systems (see Chapters 3–5). A dream is a construct of mind that includes biological components. Biological constructs, even those as important to dreaming as REMS and representational memory, are clearly limited in their capacity to explain this complex state. The opportunities for further phenomenological investigation are boundless.

THE SHADOW PLAY OF CONSCIOUSNESS

Dreams are nothing more or less than what we remember of the cognition that takes place in our own brains during sleep. An ancient sage, one of the writers of the Dead Sea Scrolls, once woke in the morning to write, "Who so regardeth dreams is like him that catcheth at a shadow, and followeth after the wind" (*Ecclesiasticus* 34.2) (58). He, like us, was a dreamer. But he was not like me, a dreamer moved by the aforementioned underground stones. He was more akin to the sailor who relies on the movement of the wind. And like some modern neurobiological theorists, his quote suggests that dreams can have no meaning beyond the shadow play of consciousness. Yet his written legacy considers the reflexive nature of inner consciousness and hints at the hidden power of dreams. He is a dreamer inspired to ask major questions, attempting to differentiate the true from the false, oneself from the self of others, and the real from the unreal. It is through dreams that we have the possibility of following the wind that drives us forward both as individuals and as a species – catching at shadows, the powerful and creative shadows that are our dreams.

Notes

1. Begouen, R. and Clottes, J. (1981) Apports mobiliers dans les cavernes du Volp (Enlene Les Trois-Feres, Le Tuc d'Audoubert) Altamira Symposium. In the art history literature, there is much debate as to whether these are actual quotes. Picasso did visit the caves, and saw the paintings at a time when his every utterance was dissected and potentially embellished or even created *de novo*. But "None of us could paint like that" is a great quote. Another quote also attributed to Picasso from the same visit, "We have learned nothing in 12,000 years," for the purpose of this presentation, may even be better (Curtis, G. [2006] *The Cave Painters*. New York: Anchor Books, p. 96). At the time of Picasso's visit, with pre-carbon dating, this was the presumed age of the cave paintings.
2. Levi-Strauss, C. (1968) *Structural Anthropology*, trans. C. Jacobson and B. Grundfest Schoepf. London: Penguin.
3. Curtis, G. (2006) *The Cave Painters: Probing the Mysteries of the World's Finest Artists*. New York: Anchor Books. Dates are based on dating of carbon pigments and remnants of site fires.
4. Lewis-Williams, J. (2002) *The Mind in the Cave: Consciousness and the Origins of Art*. London: Thames and Hudson.
5. Krings, M., Stone, A., Schmitz, R. W., Krainitzki, H., Stoneking, M. and Pääbo, S. (1997) Neandertal DNA sequences and the origin of modern humans. *Cell* 90: 19–30. Ovchinnikov, I. V., Götherström, A., Romanova, G. P., Kharitonov, V. M., Lidén, K. and Goodwin, W. (2000) Molecular analysis of Neanderthal DNA from the northern Caucasus. *Nature* 404: 490–493. While this work supports this hypothesis, much debate persists.
6. Lewis-Williams, J. (2002) *The Mind in the Cave*; Curtis, G. (2006) *The Cave Painters*. Even those supporting a major role for *Homo neanderthalensis* in prehistory admit that the overwhelming majority of evidence indicates that these cave paintings were created by groups of ancestor *Homo sapiens*.
7. Wynn, T. and Coolidge, F. L. (2012) *How to Think Like a Neanderthal*. New York: Oxford University Press, p. 119.
8. Curtis, G. (2006) *The Cave Painters*, p. 37. One proposal has been that the paintings were created in this area because that this was one of the few coexisting and overlying areas of occupation by both ancestor *Homo sapiens* and *Homo neanderthalensis*.
9. Finlayson, C. (2009) *The Humans Who Went Extinct: Why Neanderthals Died Out and We Survived*. Oxford: Oxford University Press.

10. Finlayson, C. (2009) *The Humans Who Went Extinct*, p. 217.
11. Edelman, G. (1990) *The Remembered Present: A Biological Theory of Consciousness*. New York: Basic Books.
12. Humphrey, N. (1998) Cave art, autism, and the evolution of the human mind. *Cambridge Archaeological Journal* 8: 165–191.
13. Raphael, M. (1945) *Prehistoric Cave Paintings*, trans. N. Guterman. The Bollingen Series 4. New York: Pantheon. Although repeated as a theoretical perspective many times, as discussed in Curtis's *The Cave Painters* (p. 11), the art historian Max Raphael was the first to make this supposition in print.
14. Lewis-Williams, J. (2002) *The Mind in the Cave*.
15. States, B. (1997) *Seeing in the Dark: Reflection on Dreams and Dreaming*. New Haven, CT: Yale University Press, p. 97.
16. Kosslyn, S. (1994) *Image and Brain – The Resolution of the Imagery Debate*. Cambridge, MA: MIT Press.
17. Reinach, S. (1899) Gabriel de Mortillet. *Revue Historique* 69: 67–95. Discussed in Lewis-Williams, J. (2002) *The Mind in the Cave*.
18. Laming-Emperaire, A. (1959) *Lascaux: Paintings and Engraving*, trans. E. F. Armstrong. Baltimore, MD: Penguin Books. Her life and contributions to the study of the cave art are discussed in detail by Curtis, G. (2006) *The Cave Painters*, p. 140.
19. Lewis-Williams, J. (2002) *The Mind in the Cave*; Finlayson, C. (2009) *The Humans Who Went Extinct*.
20. Curtis, G. (2006) *The Cave Painters*, p. 19.
21. Curtis, G. (2006) *The Cave Painters*, pp. 18–20.
22. Clottes, J. and Lewis-Williams, D. (1998) *The Shamans of Prehistory: Trance and Magic in the Painted Caves*. New York: Harry N. Abrams.
23. Buckley, K. (2009) *Dreaming and the World's Religions*. New York: New York University Press. Chapter 4.
24. Kalweit, H. (1988) *Dreamtime and Inner Space: The World of the Shaman*. Boston, MA: Shambhala, Chapter 13, pp. 160–174; Curtis, G. (2006) *The Cave Painters*, Chapter 4.
25. Pagel, J. F. and Vann, B. (1992) The effects of dreaming on awake behavior. *Dreaming* 2: 229–237. Pagel, J. F. and Vann, B. (1993) Cross-Cultural Dream Use in Hawaii. *Hawaii Medical Journal* 52(2): 44–45.
26. Buckley, K. (2009) *Dreaming and the World's Religions*.
27. Murzyn, E. (2008) Do we only dream in colour? A comparison of reported dream colour in younger and older adults with different experiences of black and white media. *Conscious Cognition* 17: 1228–1237.
28. Hobson, J. (1999) *Dreaming as Delirium*. Cambridge, MA: MIT Press.
29. Oppenheim, A. L. (1956) The interpretation of dreams in the ancient near east with a translation of an Assyrian dream book. *Transactions of the American Philosophical Society* 43: 179–343.
30. Kriwaczek, P. (2010) *Babylon: Mesopotamia and the Birth of Civilization*. New York: Thomas Dunn Books, pp. 132–133.
31. Buckley, K. (2009) *Dreaming and the World's Religions*, Chapter 4.
32. Lewis-Williams, J. (2002) *The Mind in the Cave*.
33. Wynn, T. and Coolidge, F. L. (2012) *How to Think Like a Neanderthal*, pp. 71–72.
34. Fishbein, H. (1976) *Evolution Development and Children's Learning*. Pacific Palisades, CA: Goodyear, p. 8; adapted by Moffatt, A. (1993) Introduction, in *The Functions of Dreaming*, ed. A. Moffitt, M. Kramer and R. Hoffmann. Albany, NY: SUNY Press, pp. 2–3.
35. Moffatt, A. (1993) Introduction, in *The Functions of Dreaming*.
36. Hofstadter, D. R. (1985) *Metamagical Themas*. New York: Basic Books. Making variations on a theme can be considered the crux of creative thought. Barrett, D. (2001) *The Committee of Sleep: How Artists, Scientists and Athletes Use their Dreams for Creative Problem Solving*. New York: Crown/Random House.

37. Hartmann, E. (1991) *Boundaries in the Mind: A New Psychology of Personality*. New York: Basic Books. Kramer, M. (2010) Dream differences in psychiatric patients, in *Sleep and Mental Illness*, ed. S. R. Pandi-Perumal and M. Kramer. Cambridge: Cambridge University Press, pp. 375–382.
38. Melville, H. (1851) *Moby-Dick: Or the Whale*. Berkeley, CA: University of California Press (reprint 1979).
39. *Webster's New Universal Unabridged Dictionary* (1996) New York: Barnes and Noble.
40. *Random House Dictionary of the English Language* – 2nd ed. Unabridged (1989). New York: Random House.
41. Rogers, C. R. (1954) Toward a theory of creativity, etc.: A review of general semantics, Vol. 11, reprinted in H. H. Anderson (ed.), *Creativity and its Cultivation*. New York: Harper and Row.
42. Guilford, J. P. (1967) *The Nature of Human Intelligence*. New York: McGraw-Hill. Creative problem solving is often based on one's ability to go off in different directions when faced with a problem.
43. Lee, H. (1940) A theory concerning free creation in the inventive arts. *Psychiatry* 3. From a "self-expressive" perspective, creativity has been defined as "the ability to think in uncharted waters without influence from conventions set up by past practices."
44. Ghiselin, B. (1952) *The Creative Process*. Berkeley, CA: University of California Press. Creativity is the process of change, of development, of evolution, in the organization of subjective life.
45. Pagel, J. F. and Kwiatkowski, C. F. (2003) Creativity and dreaming: Correlation of reported dream incorporation into awake behavior with level and type of creative interest. *Creativity Research Journal* 15: 199–205. Included data are based on this paper.
46. Delaney, G. (1979) *Living Your Dreams*. San Francisco, CA: Harper Books.
47. Wallas, G. (1926) *The Art of Thought*. New York: Harcourt, Brace.
48. Barrett, D. (2001) *The Committee of Sleep*.
49. McGinn, C. (2004) *Mindsight: Image, Dream, Meaning*. Cambridge, MA: Harvard University Press, pp. 151, 159.
50. Pagel, J. F. and Kwiatkowski, C. F. (2003) Creativity and dreaming.
51. Pagel, J. F., Kwiatkowski, C. and Broyles, K. (1999) Dream use in film making. *Dreaming* 9: 247–296.
52. Pagel, J. F. (2003) Non-dreamers. *Sleep Medicine* 4: 235–241. Included data are based on this paper.
53. Buckley, K. (2009) *Dreaming and the World's Religions*, p. 86.
54. Gillespie, G. (2002) Dreams and dreamless sleep. *Dreaming* 12: 199–208. Pagel, J. F. (October 2012) Personal report: Interview with Xin Yuan, Vice Abbott, Jingza Monastery, China. "Is dreamless sleep," I asked, "close to the experience of enlightenment?" He noted that as monks became more proficient at meditation, many reached the point at which they slept very little. With time, most no longer dreamed. In follow-up, "Sir, do you dream?" His answer, "I no longer dream."
55. Wynn, T. and Coolidge, F. L. (2012) *How to Think Like a Neanderthal*, pp. 158–160.
56. Freud, S. (1916/1951) *Beyond the Pleasure Principle*, Vol. 18, The Standard Edition, trans. and ed. J. Strachey. London: Hogarth.
57. Tootell, R., Silverman, M., Switkes, E. and De Valois, R. (1982) Deoxyglucose analysis of retinotopic organization in primate striate cortex. *Science* 218: 902–904. Felleman, D. and Van Essen, D. (1991) Distributed hierarchical processing in primate cerebral cortex. *Cerebral Cortex* 1: 1–47.
58. *Ecclesiasticus* 34.2, from *The Inclusive Hebrew Scriptures*, Vol. III (2000). Lanham, MD: Altimara Press.

Shamans, Dreams, and Religion

... Injections of that sort should not be made thoughtlessly ... and probably the syringe had not been clean. **(Freud, 1900) (1)**

Sitting on a bench in a cave, unable to move our heads, we stare at the moving shadows on the wall, convinced that these shadows are the real world of our senses. Breaking free of shackles, the one that leaves the cave is blinded by the sunlight of the exterior world. Gradually, he or she begins to see that the projected shadows on the cave's wall had been nothing more than illusions. And what if that individual chooses to return to the cave and attempts to tell this fantastic story? Would that individual be lauded, excluded, or perhaps even committed and chained for the treatment of insanity? Plato's Cave, as recorded in the *Republic*, is the parable that lies at the basis of our Western philosophy of consciousness (2). This parable has had persistent modern resonance for philosophy, literature, and even popular song (3). This parable is also a story of dreaming and its relationship to shamans, painted caves, and the consciousness of the sleep/underworld (4).

Since their discovery, there has been a series of theories as to the social role and function of the Paleolithic cave art sites. Today, it is generally accepted that the caves functioned as sites for the practice of mysticism, magic, and shamanism. The painting sites can be described and explained based on their potential roles in shamanistic ritual. The structure and design of the sites support this theory. Many sites include an audience chamber and either one or a series of passages leading deeper into the darkness, some of which terminate in remote parts of the cave on arresting images in which human figures sometimes morph into animal forms (5).

Shamanism, tightly defined, is a religious phenomenon that first developed 5000 years ago in Siberia and Central Asia. In shamanistic societies, the magicoreligious life centers around the role of a magician and medicine man called a "shaman." Based on studies of modern societies still practicing shamanism in this strict sense, most of the peoples incorporating this mysticism/religion are nomadic to some degree. The economies of these societies are most often based on hunting–fishing or herding–breeding. Although initially considered to be characteristic of Siberian and northern European societies, historic shamanistic societies were actually circumpolar, extending into Greenland and North America (6).

More loosely defined, shamans can be considered to be society's magicians, sorcerers, medicine men, and ecstatics. Such individuals are found throughout ancient and modern religions and can serve as a basis for religious ethnology (7). The views and perspectives of shamans have been incorporated into all of our major religious texts (8). A primary characteristic of the shamanistic worldview, both tightly and loosely defined, is the incorporation of altered states of consciousness into the mystical/religious process. The mystic, magic, and "religious" aspect of shamanism has been characterized as, "… the archaic technique of ecstasy" (9). In historical shamanistic societies, these episodic periods of ecstasy are generally described as experiences in which the "soul" leaves the body (10). In most traditional societies this ecstatic aspect is obtained via dreams, or through drug-, sickness-, or religion-induced trance states. Except among those who inherit their position, an ecstatic experience is almost universally a part of an individual's initiation into the shaman role. In some cultures, the ecstatic experience is described as a descent into the otherworld for conversations with dead spirits and the souls of dead shamans (11). The ecstatic experience is coupled with training in ritual by other shamans and elders in the clan in shamanistic techniques, the names and functions of spirits, the mythology, and the genealogy of the clan. Shamans are often believed to have the power to transform themselves into animals, kill at a distance, foretell the future, and perform all manner of magical tricks (12). Particularly in North America, the granting of shamanistic powers comes after deliberate questing involving travel and fasting in solitary places, most often mountain caves (13). Shamans may use music and drumming in their rituals (14). For some cultures, including the San of South Africa and the Aborigines of Australia, pictorial rock art is incorporated into shamanistic dances and rituals (15). In some of the historically described shamanistic societies, drugs, psychiatric and medical illness (particularly epilepsy), and religiously induced trances are used in the induction of ecstatic experiences (16).

Many early human communities included at least one individual who served as a healer, ritual specialist, and mediator between the living and the dead (17). The social effects of shamanism are not restricted to mysticism and religion. In tribal societies, the role of shaman is as much a political as a religious one, with the shaman functioning in the society side by side with the chief or occasionally merged into a single role (18). However, in most tribal societies, the primary role for the shaman is as a medicine man in healing and the curing of illness (19). Prescientifically, illness was generally considered a manifestation of unseen subtle malignant forces, amenable to the cures and healing of shamanistic rituals and rites.

In tribal hunter–gathering societies, shamanism, loosely defined, is a nearly universal phenomenon described historically in tribal groupings in South Africa, Australia, North and South America, and the Pacific (20). If, as evidence suggests, the 35,000-year-old painted caves are shamanistic sites, the shamanistic worldview is likely to pre-date and contribute to all other known mystical systems of belief and religion. While simple conceptually, this perspective is not clearly supported by the archaeological evidence. Tightly defined Siberian shamanism developed relatively recently among the Bronze Age hunters and herders of the far north, and there is no evidence for Siberian shamanism having a history dating back more than 5000 years (21). The historic and ethnographic data on shamanism must therefore be integrated cautiously with the apparent shamanistic aspects of the cave painting sites. Those sites date from the deep prehistoric past. The archaeological prehistory needs to be understood, as far as possible, in its own right, and not by imposing the ways of life of recent peoples on those ancient cultures. However, there is strong archaeological evidence describing the social, shamanistic, and most likely religious use of the Paleolithic cave painting sites. The caves include anterooms where gatherings took place and special paintings hidden in deep recesses that can barely be approached. There are altars of piled cave bear skulls set before paintings, the broken remains of decorated oil lamps, and musical relics including flutes (22). This evidence suggests that the painted caves functioned as archaic shamanistic sites. If so, at these sites, the paintings were likely incorporated through a shamanistic worldview into the healing, politics, and religion of the society.

BOX 2.1

EVIDENCE SUMMARY: THE PAINTED CAVES OF WESTERN EUROPE WERE ANCIENT SHAMANISTIC SITES

- Cave site design is likely to reflect the Paleolithic understanding of the nether world: darkness requiring artificial lighting with paintings that appear to move in this light, entrance chambers and deeper passages leading to panels of the art that incorporate the rocks and protrusions of the caves into its motif.
- The paintings are high-quality, non-naturalistic, and multidimensional; they incorporate multiple perspectives, apparent motion, and storylines – artistic qualities that suggest incorporation of altered states of consciousness in their creation.
- Apparent ritual use of the sites is present, including offerings, altars, and presumptive evidence for musical accompaniment.
- The long existence and development of the sites suggest that they were part of a cultural/religious focus and were cultural sites of pilgrimage.
- At their farthest and darkest reaches, the paintings include images of shamanistic themes including human–animal transformations.
- Historically persistent hunter–gathering cultures often have a shamanistic focus that may include ritual and practice involving cave art and paintings.

Despite considerable ongoing debate, this thesis is generally accepted. Evidence is presented in far greater detail and extent by authors focused on this area of study, including Eliade, Lewis-Williams, Pearce, and Whitley. Readers with an interest in further and deeper understanding of the painted caves should start with these texts and others referred to in this chapter.

This evidence, in summation, suggests that the cave paintings were incorporated into an ancient shamanistic worldview that included the role of shaman in the society, belief systems, and rituals associated with the cave painting sites (Box 2.1). Of course, any attempt to understand the paintings and/or the Paleolithic mind must be considered theoretical, but this theory of ancient shamanism has excellent external and internal consistency with the objective archaeological and ethological data. Based on current evidence, it may actually be true.

VERY DISTURBING DREAMS

The cave paintings must have exerted a powerful influence on the culture of our ancestors. While restricted to this small area of southern Europe, they are some of the most important primary evidence portraying the working of the painters' minds. The paintings illustrate basic and important characteristics of our ancestors. Being so basic, and occurring so early in the development of the modern human, those characteristics are likely to still be characteristic of our species.

In many ways the cultural role of shaman, as portrayed by the likely function of the cave paintings in rituals, defines what we know of the religious and sociopolitical behavior of our ancestors. Today, shamanism, tightly defined, still exists, particularly among native and aboriginal tribal populations. In many areas of the world, such societies are accorded marginal status and subject to discrimination and isolation. Members who maintain traditional approaches and practice are often labeled as primitive. Scientific attitudes toward these groups can be condescending, with biologists perceiving them as evolutionary relics, sociologists as cultural anomalies, psychologists as prerational, and economists as underdeveloped societies needing restructuring. To military strategists they are non-existent. To the Western tourist, such peoples may seem romantic and exotic (23). Shamanism, more loosely defined, has persisting effects even on our modern Western society. There are many alternative health practitioners who, in part, assume the role of shaman. Even in the halls of mainstream Western medicine, it is the rare patient who does not see his allopathically trained medical subspecialist as at some level a "healer." And it is the rare physician who is able to avoid that role and practice only scientific evidence-based medicine. The ancient shamanistic roles of mystic, healer, artist, and magician have persisted into our modern culture. As the descendants of the cave painters, we have preserved this role. Our shamans rarely call themselves shamans these days. They are our politicians, actors, and artists, our priests and our physicians.

In modern society, just as in "primitive" shamanistic societies, a primary role for a shaman is in the treatment of disease and healing. In the fourth century BC in Greece, this approach was incorporated into what eventually became the science of medicine. There, physician–priests of Aesculapius required their patients to sleep in the temples at the foot of the statue of their god, "incubating" a dream that was to be reported the next morning. This shamanistic approach to healing used the patient's dreaming to determine medical diagnosis and treatment. The physician's task was to interpret that dream and prescribe a cure based on his interpretation of the dream's significance. Today, the ruins of these temples are littered with stone tablets describing recorded dreams and the subsequent attempts at cure. These are historical examples of the early scientific method as applied to medicine in which symptoms, dreams, successful and failed attempts at treatment were recorded for further reference. Even surgical instruments have been recovered from temple excavations, the products of what may have been very disturbing dreams. The successes of the physician–priests that have been recorded include the surgical removal of the bladder stones, and the use of the trepanome – a hole bored through the skull to relieve cranial pressure – for the treatment of debilitating headaches resulting from intracranial bleeding. These same approaches are still used on patients with similar signs and symptoms. The most renowned of ancient physicians, including Hippocrates and Galen, were initially trained to be priests in the cults of Aesculapius (24). Shamanistic dream interpretation used in the treatment of illness led to what we call medicine today. It was only when Hippocratic physicians begin to categorize illness based on causal explanation, observation, division, and explanation that diseases were no longer viewed as the concrete effects of deliberate divine dispensation or god-sent pollutions that could be treated only by a shaman or priest (25).

DREAMS AS DIVINE EPIPHANY

The role for shaman as healer is ancient and deeply ingrained into both our society and our consciousness. It is, however, the cultural role for shaman as priest or preacher that has had an even greater and profound effect on human society. Priest–prophet–shamans developed and disseminated the religious beliefs that define our behavior and our history as a species. Philosophically, these concepts are complex and often poorly defined. The term "religion" is generally used to describe an awareness of powers that transcend human control and understanding, and have a formative influence on and an active presence within human life (26). Religion has at least three interlocking dimensions: religious experience that includes transcendent and euphoric states; religious beliefs that develop from attempts to codify these experiences in a social context; and religious practice that includes the rituals designed to manifest the belief system (27). Every religion includes aspects of ancient shamanism that have contributed to the development and structure of the belief system. Since most gods are not writers, all the major religious texts include the writings of prophets, mystics, and sages who wrote, at least in part, out of states of ecstatic communion with and understanding of their gods. These texts are among the historically oldest of human records. They rarely include evidence of drug- or trance-induced writings, but they do include a remarkable number of dreams. From the study of these texts, it is clear that humans have always expressed, discussed, and recorded their dreams, and then used them to support and inspire systems of belief. From the beginning of recorded history, dreaming has been regarded as a religious phenomenon (28).

The Hindu Vedas, written down some time between 1200 and 900 BCE, are probably the oldest extant writings of an active modern religion. In general, these Hindu traditions view dreams as products of the human mind, rather than transmissions from the gods. In some of the oldest texts collected in the *Rig Veda*, the view of dreams relates to their shamanistic use in curses:

> We have conquered today, and we have won; we have become free of sin. The waking dream, the evil intent – let it fall upon the one we hate; let it fall upon the one who hates us. *(Rig Veda) (29)*

The Vedas also include descriptions of favorable and unfavorable dreams, as well as speculations concerning the expected realization of prophecies based on the time of night of the dream experience.

In China, a historic use of dreams in tightly defined shamanism preceded the development of Confucianism. Dreams, perceived as communications from dead ancestors, were used in prophecy. The Chinese Imperial Courts had official roles for dream interpreters and divination specialists. Modes and directions for the use of dreams in divination were recorded in the Zhou-era (1100 BCE) Confucian classic *Shi-jing, The Book of Odes* (30).

Life stories of the Buddha include numerous dreams. This suggests that dreams had a prominent role in that society as conducive agents for spiritual insight. In early Buddhist texts (500 BCE), the Buddha's birth is heralded by a dream of divine conception, a subgenre of religious dreams also present in earlier Jainistic literature and later Christian texts (31). In the New Testament, Joseph was told in dream of the miraculous basis of Mary's pregnancy and that the child should be named Jesus (32). Such dreams served to support the legitimacy of new religious leaders and their followers in their attempts to displace the existing dominant religious hierarchies.

In the *Torah*, God famously proclaims, "If any one among you is a prophet, I will make myself known to him in a vision, I will speak to him in a dream" (33). The stories of Abraham and his immediate descendants include many highly significant dreams perceived as primal sources for divine inspiration (34). Jacob's dream is one of the most powerful passages of the Old Testament, a text incorporated into theology and used as justification for world conquest by at least three of the world's major religions.

> [Jacob] came to a certain place, and stayed there that night, because the sun had set. Taking one of the stones of the place, he put it under his head and lay down in that place to sleep. And he dreamed that there was a ramp set upon the earth, and the top of it reached to heaven; and behold the angels of God were ascending and descending on it. And behold, the Lord stood above it and said, "I am the Lord, the God of Abraham your father and the God of Isaac; the land on which you lie I will give to you and your descendants; and your descendants shall be like the dust of the earth, and you shall spread abroad to the west and to the east and to the north and to the south; and by you shall all the families of the earth bless themselves. Behold, I am with you and will keep you wherever you go. And will bring you back to this land: for I will not leave you until I have done that of which I have spoken to you." Then Jacob awoke from his sleep and said, "Surely the Lord is in this place; and I did not know it." And he was afraid and said, "How awesome is this place! This is none other than the house of God and this is the gate of Heaven." *(Genesis 28) (35)*

Joseph in Egypt explained the Pharaoh's dreams, and suggested that faith in one's god is what gave an individual the divine ability to interpret the dreams of others (36). Moses had a more skeptical view of dreaming:

> If a prophet arises among you, or a dreamer of dreams, and gives you a sign or a wonder, and the sign or wonder which he tells you comes to pass, and if he says, "Let us go after other gods," which you have not known, "and let us serve thee," you shall not listen to the words of that prophet or to that dreamer of dreams; for the Lord your God is testing you, to know whether you love the Lord your God with all your heart and all your soul .... That prophet or that dreamer of dreams shall be put to death, because he has taught rebellion against the Lord your God. *(Numbers 12) (37)*

This approach to dreaming has been incorporated into aspects of Christianity in which dreams are considered as temptations from the devil liable to lead the dreamer astray. This perspective became predominant during the inquisitions of the Middle Ages and the American witch trials of the seventeenth century, and was used as a theological basis supporting the executions of witches and dream interpreters.

> Those things, whether past, present, or to come, which ... cannot be known by human skill in arts or strength of reason ... nor made known by divine revelation ... must needs be known (if at all) by information from the devil. *(Rev. Cotton Mather, in Demos, 2008) (38)*

Recently, such theological pronouncements regarding "false" dream interpretation provided justification and support for the *fatwā* ordered against Salman Rushdie for his reinterpretation of the dreams of Muhammad in the book *The Satanic Verses* (39).

Despite such perspectives in their ancestral traditions, later prophets continued to interpret, use, and record their dreams in their religious texts. During exile in Babylon, Daniel was able to gain political power as a dream interpreter for Nebuchadnezzar:

You saw, O King, and behold a great image. This image mighty and of exceeding brightness, stood before you, and its appearance was frightening. The head of the image was fine gold, its breast and its arms of silver, its belly and thighs of bronze, its legs of iron, its feet partly of iron and partly of clay. As you looked, a stone was cut out by no human hand, and it smote the image on its feet of iron and clay, and broke them in pieces; then the iron, the clay, the bronze, the silver, and the gold, all together were broken in pieces, and became like the chaff of the summer threshing floors; and the wind carried them away, so that not a trace of them could be found. But the stone that struck the image became a great mountain and filled the whole earth. *(Daniel 2) (40)*

It seems as if Daniel was cognizant with the dream record of King Gudea (Chapter 1, pp. 7–8). Many of the later prophets included dreams in text. The Berachot, Chapters 55–57 of the *Talmud*, focuses on the use and interpretation of dreams (41). Much of the *Qur'an* (*Koran*) was revealed to Muhammad in dreams, and dream visitations from the Angel Moroni led Joseph Smith to find and reveal the Book of Mormon (42).

In Judeo-Christian tradition, the perspective exists that direct dream prophecy may have ended with the writing of the *Bible*. However, dreams continue to have roles in religious worship and in achieving religious understanding. This was perhaps best explained by Rabbi Solomon Almoli in his sixteenth century *Pitron Chalomot*: "Since the time of our exile from our homeland Israel, prophecy came to an end and the oracle was hidden from us. Yet even so, we have retained our ability to be inspired by dreams, which tell us of all that will come to pass" (43). Most modern religions have sects that accept that dreams can provide direct assess to their deity. In the Eastern Orthodox tradition, such dreams are referred to as "manifestations of the divine." Evangelicals and Pentecostals may refer to such dreams as the "prophetic word of the Lord" and communications from the Holy Spirit. Many Catholic and Protestant sects believe that dreams can provide a "divine epiphany," an awareness of the "*mysterium tremendum*" of the holy presence. Some theologians have even adapted psychoanalytic dream theory in the attempt to use dreams in understanding biblical scripture (44). Mormons speak of the "direct revelation" available in prophetic dreams.

Hindus refer to "*darshan*," a glimpse or vision of the divine that can occur in dreams. Muslims use a form of dream incubation called "*istikhara*," focusing on consulting God and asking for the blessing of his knowledge and guidance (45).

Within such traditions lie the possibility for shamanistic dreams. Even today, kings, prophets, and priests sometimes proclaim that they have had dreams that are messages from God. Such theocratic dreams can have profound effects on people's lives and world events. In 1994, the Taliban leader Mullah Omar announced that Muhammad had appeared to him in a dream and instructed him to take action to save Afghanistan from corruption and foreign powers (46) (Box 2.2).

BOX 2.2

EVIDENCE SUMMARY: SHAMANISM PERSISTS TODAY AS A MAJOR PSYCHOLOGICAL AND SOCIAL FACTOR AFFECTING OUR WAKING BEHAVIOR

- The cave paintings that probably reflect ancient shamanistic belief and practice are among the earliest recorded records of our species; such historically early behavior reflects basic species characteristics that are extremely likely to be both genetically and culturally preserved by the species.
- The shamanistic role of mystic, prophet, magician, dream interpreter, and healer contributed to all forms and texts of what we call religion, and religion continues to be pervasive in modern society.
- The shamanistic role of mystic, prophet, magician, dream interpreter, and healer led to and has been incorporated into what we today call medicine.
- Alternative approaches to health and healing beyond those utilized in scientifically based allopathic medicine continue to be popular and often successful approaches to health care.
- Our Western society maintains a focus in its literature and entertainment, including film, music, and gaming, on mythology and shaman-like protagonists who have magic and super-human capabilities.
- Alternative conscious experiences continue as a major aspect of modern society.

THE PURKINJE TREE: SHAMANISM AND TRANCE STATES

Visiting the painted caves, even when alert and awake, is a perceptually dissociating experience. The descent into the cave, the flickering of artificial lights, the pools of absolute darkness, and the incredible painted images of extinct creatures can induce an experience that resembles a drug- or illness-induced trance state, or a dream. Contrarily, the experience induced by today's above-the-ground replicas of the caves is most often devoid of quiet and awe, more reminiscent of a visit to the county fair.

In 1988 David Lewis-Williams and Thomas Dowson proposed that the wall markings and the figures depicted in the painted caves might provide a "neurological bridge" connecting us to the mental constructs of the Paleolithic painters (47). They were working from within the prevailing perspective in cognitive archaeology, the belief that increasing cranial capacity in early hominids led to expanded memory capacity and increased intelligence (48). They suggested that increased working memory capacity in early humans allowed for a recollection of their spiritual world that could be incorporated into paintings. They proposed that these recollections occurred during altered states of consciousness, and specifically that these were visionary images reflecting hallucinatory experiences. They proposed that these markings (paintings) were produced by the human nervous system during hallucinatory trance (49).

This theory suggests that mental/visual imagery during trance states progresses through three stages. In the first stage, geometric light patterns, what Lewis-Williams and Dowson called "entoptic" (within the eye) patterns, are described. Visual migraine prodromes that can include herring bone patterns and jagged lines are examples of such phenomena. Other examples include the visual auras experienced in association with some forms of epilepsy, and the visual effects described by individuals who have ingested mescaline and lysergic acid diethylamide (LSD) (50). Lewis-Williams and Dowson made an encyclopedic collection of the non-pictorial rock art present in the painted caves, and categorized these images into six principal forms: dots, zigzags, grids, nested curves, parallel lines, and filigrees. They proposed that these non-pictorial cave markings were made during this first stage of hallucinatory trance.

During the second stage of such trance states, they proposed that different visual patterns develop, essentially normal visual images that are influenced by the social and cultural milieu. During this stage, the trance subject is trying to make sense of the entoptic patterns by, for example, seeing a grid as a chessboard. And in the third and final stage, full-blown hallucinations are experienced, in which the individual may imagine becoming the thing that she or he hallucinates. Examples include various paintings purported to show the transformation of human shamans into animals. The multidimensional, multirepresentational, and self-inclusive characteristics of some of the fully developed paintings, interpreted in light of this theory, could be the result of a fully developed hallucinatory trance (51). The historic use of induced hypnogogic hallucinations in surrealistic art by painters such as Salvador Dali supports this final component of their theory (52).

There are troubling aspects to this model. It is unclear whether their theory meets accepted anthropological criteria for archaeological validity (Table 2.1). In modern society, ecstatic states are unusual even with the wide availability of varied psychotropic drugs. It is only recently that scientific methodology and techniques have been applied to the study

TABLE 2.1   Methodological Criteria for a Theory in Cognitive Anthropology (53)

Archaeological validity	The archaeological evidence must itself be credible. The traces in question must be reliably identified and placed appropriately in time and place
Cognitive validity	The evidence must actually require the abilities attributed to it. The cognitive ability must be one recognized or defined by cognitive science; it must be required for the actions cited; and the archaeological traces must require those actions. A strict standard of parsimony must apply. If the archaeological traces could have been generated by simpler actions, or simpler cognition, then the simpler explanation should be favored

of the varied altered states of consciousness. Because of a lack of clear biological markers, studies of personally experienced cognitive states require individual reports of the experience being studied. The scientific study of such states is replete with studies confounded by methodological and transference variables that confuse the apparent findings (54). Studies of psychedelic visual drug effects that date from several decades ago are often anecdotal, based on single patient reports and methodology that does not control for transference effects (55).

Very little work has been done addressing the visual imagery typical of trance states. The visual aspects of physiologically normal states such as drowsiness, focused gaze, or repetitive eye blinking are almost unstudied (56). The visual patterns that occur on morning awakening from sleep can be classified based on light patterns described during lucid dreaming. Such light patterns include lattice patterns including vertical and horizontal lines, squares, chessboard images, and continuous hexagons, and are described as occurring in a rounded circumspect field (57). Eye floaters are perhaps the most commonly experienced of entoptic phenomena: blobs of varying size, shape, and transparency, particularly noticeable when viewed against a bright, featureless background such as the sky or a point source of diffuse light very close to the eye. Floaters are the shadow images of objects suspended just above the retina. Other entoptic phenomena observed by some individuals include "blue-field phenomena": tiny bright dots moving along lines of the visual field – most likely white blood cells moving in capillaries in front of the retina (58). The Purkinje tree is a reflected image of the retinal blood vessels in one's own eye. It was first described by Purkinje in 1823 (59). These patterns fit poorly into the entoptic cave art classification system. Most are not described by individuals experiencing drug-induced trance states.

An anthropologist specializing in non-pictorial cave markings can classify those markings according to the Lewis-Williams and Dowson system, as can a neurologist confronted with visual migraine pro-dromes or seizure-associated hallucinations (60). Entoptic visualizations were sometimes used by surrealist artists. A technique called "entop-tic graphomania" was developed as an automatic method of drawing in which dots were made at the sites of impurities in a blank sheet of paper, and lines are then made between the dots, with these being either "curved lines … or straight lines" (61). Graffiti and schoolroom doodling, today's versions of cave painting, incorporate non-pictorial patterns that, subjected to a pattern analysis, would be likely to produce similar find-ings. The entoptic imagery model is based on very limited empirical evi-dence for the association of these light patterns with trance states. The authors' classification system for these light patterns probably reflects their extensive background in studying non-pictorial cave art patterns. It is tempting to suggest that many hours spent studying wall scrapings in dark, wet, poorly accessible cave sites might lead to an altered state of consciousness, in which the patterns of such visual experiences would be more likely to become apparent to the focused researcher. Pending fur-ther evidence, the entoptic phenomena component of this theory must be considered speculative.

But the most deflating aspect for this model is the incredible quality of the cave art. The techniques used and the representations attained by their ancient painters are far beyond what might be expected from drug-, epilepsy-, or religiousritual-induced flights of hallucinatory ecstasy. It is likely that the factors contributing to high-quality artwork in the Paleolithic era are similar to the factors contributing to the highest qual-ity artwork of today. Training in style, techniques, and perspective would be required, as well as a background in extant paintings and rock art. The Paleolithic version of art history must have required travel, spelunk-ing, and political as well as survival skills. Factors contributing to cre-ativity were likely to be similar to those required today. Drugs, disease, and ecstatic trance experience are low on the list of contributory factors to great art. It is likely that the paintings were executed during highly focused waking. The varied representations and perspectives incorpor-ated in these paintings probably came in part from an altered state but it was likely to be the same altered state (dreaming) that is commonly uti-lized in the creative art of today (see Chapter 1).

The Lewis-Williams and Dowson neurophysiological theory argues that the cave art was produced during ecstatic drug- or religious ritual-induced trance. Others have suggested a potential role for the medical diagnoses of epilepsy and hypnogogic hallucinations in the art. And oth-ers have argued that the shamans creating the art were likely to have suffered from bipolar disorder – a psychiatric disorder sometimes associ-ated with creative personalities (62). While these authors have suggested that dreams could have been one of the alternative states available to the shaman/painter, there is good evidence suggesting that rather than drugs and/or illness, dreams were the primary source of the "ecstatic" component of the cave painters' art. While dreams were not the only alternative cognitive perspectives that could be used in the creation of the cave art, just as they are for artists today, dreams were an easily available source of ecstatic creative inspiration for the cave art painters. Dreaming certainly provides a simpler explanation as a source for artistic ecstasy than the invoking of the complex multitiered experience of drug-induced trance (Box 2.3).

BOX 2.3

EVIDENCE SUMMARY: ARGUMENT AGAINST DRUG-INDUCED ECSTATIC TRANCE AND MEDICAL/PSYCHIATRIC ILLNESS BEING THE BASIS FOR THE ECSTATIC COMPONENT OF THE CAVE ART

- The theory that the non-pictorial markings on the painted cave walls are evidence for the first stage of ecstatic trance must be considered speculative.
- The complex technique and extremely high quality of the cave painting art suggests that these paintings were made deliberately and with great skill, rather than splashed on the wall while in the midst of an ecstatic trance.
- Cave art images are stylized renditions with the same styles of presentation used at multiple sites over an extended period (i.e., the cave painters must have been trained in such style as well as technique). Again, this suggests the paintings were deliberate, considered works of art rather than ecstatic daubing.
- Evidence suggests that some of the best known cave art panels were drawn in stages that involved scaffolding, markings followed by wall scraping, and later completion and filling in of the figure panels – a process that required supplying and lighting assistance over an extended rather than short ecstatic period.
- Historically, drug-induced trance and medical and psychiatric illnesses are not often a basis utilized in the creative process of artists.
- It would be expected that archaic and species-defining characteristics would persist. Today, the experience of ecstatic trance cannot be considered a major structural component of human society.

DIRTY NEEDLES AND DREAM ECSTASY

In modern shamanistic societies, shamans use their dreams and the dreams of their patients to access an altered state of consciousness. Dreams are used even by those shamans who routinely utilize psychoactive drugs, fasting, or religious ecstasy (63).

The dreams of sleep provide a nearly universally available access to an alternative state of consciousness. Whether utilized by the creative artist, actor, or scientist, dreams – even nightmares – are very often an intrinsic part of the creative artistic experience (Box 2.4).

BOX 2.4

EVIDENCE SUMMARY: DREAMS WERE LIKELY PRIMARY SOURCES FOR THE "ECSTATIC" COMPONENT OF THE CAVE PAINTER ART

- Cave design and location underground suggest that the Paleolithic understanding of the nether world may be as a descent into that strange and perceptually dissociated world of sleep.
- Since almost everyone dreams, this alternative source of content and perspective is more widely accessible than the drugs that may have been required to induce ecstatic trance.
- The complex and generally complete rendition of the cave images is most consistent with the highly developed and multiply associated images from dreaming, rather than the aura and or flashes potentially associated with illness and drug-induced trance.
- Dreaming, rather than drug-induced trance or medical and psychiatric illness, has been historically incorporated into the shaman-based roles of healer/physician, priest, and magician/politician.
- Dreaming is even today consistently utilized in the creative process of artists (this was discussed in Chapter 1).
- It would be expected that archaic and species-defining characteristics would persist, and today there are still many individuals who access and use their dreams in their daily lives, while there are fewer who routinely experience ecstatic trance.
- The study of alternative cognitive states, particularly the study of dreaming, can be difficult since the researcher or interested scientist seems very often moved to fall into an aspect of the shaman's role.

When you investigate dreaming you are also studying the role that the shaman has assumed in our society for thousands of years. To repeat the David Foulkes quote, "… There is something about dreaming that has always seemed to move people prematurely to forsake the possibility of disciplined empirical analysis" (64). It is clear that studying dreams increases one's tendency to recall dreams and opens up an altered state of conscious that differs markedly from our daily reality of focused perceptual waking. Such altered consciousness is tied culturally to the role of shaman, an authoritative social position as elder, healer, and priest that persists in our psyche. My answer to Foulkes, anecdotal as it may be, is that it is the socially attractive authoritarian role of shaman that so easily inspires the dream scientist to move off-target. It is hard to function as a shaman and at the same time maintain a logical and rational approach to evidence when studying the shadow of consciousness that we call "dreaming." It has been proposed that our dreams describe the royal road to the unconscious. It has been suggested that our dreams describe the structure of our psyche, and provide evidence for the structure and functioning of consciousness. It is far more likely that whether we are kings, priests, scientists, or shamans, the stories that we remember in the morning from our sleep are more personal. They are our dreams.

The scientist or physician who chooses to study dreaming may be particularly prone to leaning toward a "shamanistic" role. Dreams are universal and intrapersonal. Studying dreams requires that the researcher incorporate an internal approach to the subject. It would clearly be inappropriate for dream scientists to rely only on the reports of others and never look at their own dreams. Research that relies entirely on an external and applied approach is beset by methodological complications. Such researchers are more likely to make what Freud referred to as "unconscious errors" and to express what today we refer to as "Freudian slips." Dream scientists who ignore the intrapersonal aspect of dreams have misdefined the topic of study, missed transference effects, confused actual results with complicating variables, and examined their results with increasingly complex methodology and statistical analysis that sometimes hides their legitimate findings. They have often chosen to examine their data in the light of historic theory and untestable belief. Faced with such methodological complications, dream scientists frequently attempt to address the interpersonal aspects of dreaming by publishing their own dreams and/or the dreams of their patients. This approach has its own limitations. The personal dreams of dream scientists are often presented with surprisingly limited understanding of the dream's interpersonal meaning. This can happen even for individuals who have a profound understanding of the dreams of others. Perhaps the most famous of such dreams is Freud's dream of Irma's injection. This complex dream is the story of one of Freud's patients who responded poorly to his chosen approach to therapy (Box 2.5).

Freud included this dream in his seminal work *The Interpretation of Dreams* (1899), and used his interpretation of this dream (the next sixteen pages of that book) as an example of the psychoanalytic method. There have been realms of psychoanalytic literature inspired by this dream, addressing the value of psychoanalysis as a therapeutic approach. Most of this literature is written from the Freudian perspective with the dream viewed as the disguised, hallucinatory fulfillment of an infantile, often sexual wish that serves to maintain the continuity of sleep (66).

BOX 2.5

FREUD'S DREAM OF IRMA'S INJECTION (1899) (65)

Freud describes the dream, which he records as taking place on the night of 23–24 July 1895, thus:

> A large hall – numerous guests, whom we were receiving. – Among them was Irma. I at once took her to one side, as though to answer her letter and to reproach her for not having accepted my "solution" yet. I said to her: "If you still get pains, it's really only your fault." She replies: "If you only knew what pains I've got now in my throat and stomach and abdomen – it's choking me." – I was alarmed and looked at her. She looked pale and puffy. I thought to myself that after all I must be missing some organic trouble. I took her to the window and looked down her throat, and she showed signs of recalcitrance, like women with artificial dentures. I thought to myself that there was really no need for her to do that. – She then opened her mouth properly and on the right I found a big white patch; at another place I saw extensive whitish grey scabs upon some remarkable curly structures which were evidently modelled on the turbinal bones of the nose. – I at once called in Dr M, and he repeated the examination and confirmed it ... Dr M looked quite different from usual; he was very pale, he walked with a limp and his chin was clean-shaven ... My friend Otto was now standing beside her as well, and my friend Leopold was percussing her through her bodice and saying: "She has a dull area low down on the left." He also indicated that a portion of the skin on her left shoulder was infiltrated. (I noticed this, just as he did, in spite of her dress.) ... M said: "There's no doubt it's an infection, but no matter; dysentery will supervene and the toxin will be eliminated. ... We were directly aware, too, of the origin of the infection. Not long before, when she was feeling unwell, my friend Otto had given her an injection of a preparation of propyl, propyls ... propionic acid ... trimethylamin (and I saw before me the formula for this printed in heavy type) ... Injections of this sort ought not to be given so thoughtlessly ... And probably the syringe had not been clean.

The report of this dream closes as follows: "... Injections of that sort should not be made thoughtlessly ... and probably the syringe had not been clean." Freud interpreted this passage as acquitting him of the responsibility for Irma's incomplete cure: "the dream represented a particular state of affairs as I should have wished it to be. Thus its content was the fulfillment of a wish and its motive was a wish" (67). Until recently, almost all of literature avoided what today seems the obvious self-revelatory continuity of this dream and its direct comment on Freud's serious cocaine addiction (68).

Based on the evidence of the cave paintings, our species has been utilizing dreaming in the creative process for at least 32,000 years. Based on the historic record, we have been analyzing dream content for at least the past 4000 years. A comparative review of the early religious texts indicates that dreaming has contributed to and affected all of our major belief systems. Part of the difficulty with studying dreaming is that almost everyone dreams. And those with an interest in the creative process use their dreams in their work. So, on the personal level, dreaming has contributed to almost every religious, philosophical, artistic, or scientific insight. This general yet intangible pervasive effect of dreaming on personal, creative, and philosophical belief systems means that changes in our understanding of dreaming are likely to have widespread and unexpected effects. The cave paintings are a tabula rasa – exquisite Paleolithic artworks created by ancestors from tens of thousands of years ago whose minds we will never clearly understand. Because of their unique anthropological importance, as well as their beauty, they have been the object of intense theorizing and study. A changed understanding of dreaming and its biological basis affects our understanding even of these paintings, their meanings, and the evidence they provide to our history and functioning as a modern species.

Notes

1. Freud, S. (1900) *The Interpretation of Dreams*, trans. and ed. J. Strachey (1953) *Vols. IV and V of the Standard Edition*. London: Hogarth Press and Institute for Psychoanalysis; Reprint (1965). New York: Avon Press, pp. 139–140.
2. Plato. Plato's cave – the first stage, from *The Republic*, Book VII, 514a–517a, taken from *Platonis Oprea*, recogn. Ioannes Burnet. Oxonii: Clarendon (2nd ed.) (1905–1910) Vol. 4 (English trans.). Heidegger included the referenced version in *The Essence of Truth* (2002). London: Continuum.
3. Heidegger, M. (2002) *The Essence of Truth*. Saramajo, J. (2008) *The Cave*. Orlando, FL: Harcourt Books. (this book led to a Nobel Prize for literature). Mumford and Sons (2010) The cave, from *Sigh No More*, prod. M. Draves. Island Records (this song led to a Grammy Award for Album of the Year, 2012).
4. Kalweit, H. (1988) *Dreamtime & Inner Space: The World of the Shaman*. Boston, MA: Shambhala.
5. Curtis, G. (2006) *The Cave Painters: Probing the Mysteries of the World's Finest Artists*. New York: Anchor Books. Lewis-Williams, D. (2002) *The Mind in the Cave: Consciousness*

and the Origins of Art. London: Thames and Hudson. Whitley, D. (2009) *Cave Paintings and the Human Spirit: The Origin of Creativity and Belief*. Amherst, NY: Prometheus Books.

6. Price, N. (2002) *The Viking Way: Religion and Warfare in Iron Age Scandinavia*. Uppsala: Department of Archeology and Ancient History.

7. Eliade, M. (1972/1951) *Shamanism: Archaic Techniques of Ecstasy*. New York: Routledge & Kegan Paul.

8. Buckley, K. (2009) *Dreaming and the World's Religions*. New York: New York University Press.

9. Eliade, M. (1972/1951) *Shamanism: Archaic Techniques of Ecstasy*, p. xix.

10. Sullivan, L. (1994) The attributes and power of the shaman: A general description of the ecstatic care of the soul, in *Ancient Traditions: Shamanism in Central Asia and the Americas*, ed. G. Seaman and J. Day. Niwot, CO: University Press of Colorado, pp. 29–46. Stewart, O. (1994) Peyote religion, in *Ancient Traditions: Shamanism in Central Asia and the Americas*, pp. 179–186.

11. Eliade, M. (1972/1951) *Shamanism: Archaic Techniques of Ecstasy*.

12. Furst, P. (1994) Introduction: An overview of shamanism, in *Ancient Traditions: Shamanism in Central Asia and the Americas*, ed. G. Seaman and J. Day. Niwot, CO: University Press of Colorado, p. 11.

13. Park, W. (1938) *Shamanism in Western North America: A Study in Cultural Relationships*. Evanston, IL: Northwestern University Studies in the Social Sciences (2).

14. Speck, F. (1935/1977) *Naskapi: The Savage Hunters of the Labrador Peninsula*. Norman, OK: University of Oklahoma Press, pp. 178–179.

15. Lewis-Williams, D. (2002) *The Mind in the Cave*.

16. Eliade, M. (1972/1951) *Shamanism: Archaic Techniques of Ecstasy*.

17. Lewis-Williams, D. and Dowson, T. A. (1988) Signs of all times: Entoptic phenomena in upper Palaeolithic art. *Current Anthropology* 29: 201–245.

18. Buckley, K. (2009) *Dreaming and the World's Religions*.

19. Eliade, M. (1972/1951) *Shamanism: Archaic Techniques of Ecstasy*.

20. Buckley, K. (2009) *Dreaming and the World's Religions*.

21. Whitley, D. (2009) *Cave Paintings and the Human Spirit*.

22. Curtis, G. (2006) *The Cave Painters*.

23. Kalweit, H. (1988) *Dreamtime & Inner Space*.

24. Van Der Eijk, P. (2005) *Medicine and Philosophy in Classical Antiquity: Doctors and Philosophers on Nature, Soul, Health and Disease*. New York: Cambridge University Press, p. 60.

25. Van Der Eijk, P. (2005) *Medicine and Philosophy in Classical Antiquity*, p. 60.

26. Buckley, K. (2009) *Dreaming and the World's Religions*.

27. Lewis-Williams, D. and Pearce, D. (2005) *Inside the Neolithic Mind – Consciousness, Cosmos and the Realm of the Gods*. New York: Thames and Hudson.

28. Buckley, K. (2009) *Dreaming and the World's Religions*.

29. Doniger O'Flaherty, W. (trans.) (1981) *The Rig Veda*. London: Penguin Books.

30. Fang Jing Pei and Zhang Juwen (2000) *The Interpretation of Dreams in Chinese Culture*. Trombull, CT: Weatherhill.

31. Buckley, K. (2009) *Dreaming and the World's Religions*.

32. *Holy Bible. Dictionary/Concordance – Authorized King James Version. Matthew* 1. USA: Collins World.

33. *Holy Bible. Numbers* 12: 6.

34. Buckley, K. (2009) *Dreaming and the World's Religions*.

35. *Holy Bible. Genesis* 28: 11–17.

36. *Holy Bible. Genesis* 37.

37. *Holy Bible. Numbers* 12: 6–8.

38. Demos, J. (ed.) (2008) Witch-hunting in the American colonies, 1607–92, in *The Enemy Within – 2,000 Years of Witch Hunting in the Western World*. New York: Viking, p. 101.

39. Rushdie, S. (1988) *The Satanic Verses*. New York: Viking.
40. *Holy Bible. Daniel* 2 :1–34.
41. Buckley, K. (2009) *Dreaming and the World's Religions*.
42. Van de Castle, B. (1994) *Our Dreaming Mind*. New York: Ballantine Books, pp. 41, 55.
43. Almoli, S. (1500s) *Pitron Chalomot Artemidorus*.
44. Sanford, J. (1970) *The Kingdom Within*. San Francisco, CA: Harper, pp. 12–13.
45. Buckley, K. (2009) *Dreaming and the World's Religions*.
46. Edgar, I. (2004) The dream will tell: Militant Muslim dreaming in the context of traditional and contemporary Islamic dream theory. *Dreaming* 14: 21–29. Sirriyeah, E. (2011) Dream narratives of Muslims' martyrdom: Constant and changing roles past and present. *Dreaming* 21: 168–180.
47. Lewis-Williams, D. and Dowson, T. (1988) The signs of all times: Entoptic phenomena in Upper Palaeolithic art. *Current Anthropology* 34: 55–65.
48. De Beaune, S. (2009) Technical invention in the Paleolithic: What if the explanation comes from the cognitive and neurophysiological sciences? In *Cognitive Archaeology and Human Evolution*, ed. S. de Beaune, F. Coolidge and T. Wynn. Cambridge: Cambridge University Press, pp. 3–14. Rossano, M. (2009) The archaeology of consciousness, in *Cognitive Archaeology and Human Evolution*, pp. 25–35. The concept that increased cranial capacity leads to increased memory and intelligence remains the accepted belief in the field of physical anthropology. It is based on paleoanthropological data indicating that cranial capacity and frontal cortex size have increased from the first hominids to modern humans.
49. Siegel, R. (1977) Hallucinations. *Scientific American* 237: 132–140. Horowitz, M. (1964) The imagery of visual hallucinations. *Journal of Nervous and Mental Disease* 138: 513–523.
50. Kluver, H. (1966) *Mescal and Mechanisms of Hallucinations*. Chicago, IL: University of Chicago Press.
51. Lewis-Williams, D. (2002) *The Mind in the Cave*.
52. Pagel, J. F. (2008) *The Limits of Dream: A Scientific Exploration of the Mind/Brain Interface*. Oxford: Academic Press.
53. Wynn, T. and Coolidge, F. (2009) Implications of a strict standard for recognizing modern cognition, in *Cognitive Archaeology and Human Evolution*, ed. S. deBeaune, F. Coolidge and T. Wynn. Cambridge: Cambridge University Press., pp. 118–119.
54. Foulkes, D. (1993) Data constraints on theorizing about dream function, in *The Functions of Dreaming*, ed. A. Moffitt, M. Kramer and R. Hoffmann. Albany, NY: SUNY Press., pp. 11–20.
55. Krippner, S. and Kremer, J. (2010) Hypnotic-like procedures in indigenous shamanism and mediumship, in *Hypnosis and Hypnotherapy: Neuroscience, Personality and Cultural Factors*, ed. D. Barrett., Vol. 1. Santa Barbara, CA: Praeger, pp. 104–106.
56. Kosslyn, S. (1994) *Image and Brain – The Resolution of the Imagery Debate*. Cambridge, MA: MIT Press.
57. Gillespie, G. (1997) Hypnopompic images and visual dream experiences. *Dreaming* 7(3). This study can be considered a control study in which the author has had no apparent exposure to the cave art entoptic classification system, and thus created his own classification system based on concepts related to lucid dreaming. This classification system corresponds poorly with the cave art-based system of non-pictorial classification.
58. White, H. E. and Levatin, P. (1962) "Floaters" in the eye. *Scientific American* 206(6): 199–127.
59. Reid, B. (1990) Haidinger's brush. *Physics Teacher* 28: 598.
60. Claman, H. N. (2011) Visionary art? Shamans, Charles Bonnet, and the cave paintings. *The Paros*: 3–10.

61. Wikipedia. Automatic techniques. Retrieved June 21, 2008, from http://www.ithellcolquhoun.co.uk/5263/.

62. Whitley, D. (2009) Creativity and the emotional life of the shamans, in *Cave Paintings and the Human Spirit*, pp. 209–245.

63. Eliade, M. (1972/1951) *Shamanism: Archaic Techniques of Ecstasy.*

64. Foulkes, D. (1993) Data constraints on theorizing about dream function, in *The Functions of Dreaming*, ed. A. Moffitt, M. Kramer and R. Hoffmann. Albany, NY: SUNY Press, p. 18.

65. Freud, S. (1899/1965) *The Interpretation of Dreams*, trans. and ed. J. Strachey. New York: Avon Press, pp. 139–140.

66. Kramer, M. (2007) *The Dream Experience: A Systemic Exploration*. New York: Routledge, p. 149.

67. Freud, S. (1900) *The Interpretation of Dreams*, p. 151.

68. Freud, S. (1974) , in *Cocaine Papers*, ed. R. Byck. New York: New American Library. Thornton, E. M. (1983) *The Freudian Fallacy*. London: Blond & Briggs. Other reinterpretations of Freud's example dreams in light of current knowledge of the extent of his cocaine addiction include the reanalysis of his dream of the botanical monograph, as well as this dream of Irma's injection. Cole, J. (1998) Freud's dream of the botanical monograph and cocaine the wonder drug. *Dreaming* 8: 187–204.

3

Dream Philosophy

At the same moment he saw in the middle of the college court another person who called him by name in the most obliging terms, and told him that if he cared to go in search of Monsieur N. the latter had something to give him. Descartes imaged that it was a melon who had been brought to him from some foreign country. But what surprised him more was to see that those who together with this person were gathering around him for conversation stood on their feet straight and steady, whereas he himself on this same ground was still bowed and staggering, and the wind which more than once had been on the point of upsetting him had become less strong. **(Descartes, 1691) (1)**

It is not at all surprising that dreaming is of particular interest to philosophers. All of consciousness can be viewed as a form of dreaming, an illusion of our external environment as interpreted by our senses, presented in cognitively digestible form. Philosophically, dreams are more than just a report of what has occurred in the brain during sleep. Dreams are a window that can be used to see into the functioning of the mind. Dreams mysteriously occur during the perceptual isolation of sleep. That isolation makes dreams more difficult to approach, but can also be used to help in describing the limits of perceptual cognition.

Dream Science.
DOI: http://dx.doi.org/10.1016/B978-0-12-404648-1.00003-X

Philosophy has both advantages and frustrations when compared to science and religion. Philosophy emphasizes process rather than outcome. The final answers that many of us crave, while interesting and compelling, are not required. The philosopher's challenge is to ask questions, applying clarity, rigor, argument, and theory to a journey leading toward philosophical "truth." The most basic questions, while approached, are likely never to achieve total resolution. What is real, and what is dream? – Plato approached this in parable in his story of the Cave. In organizing knowledge, Aristotle applied this understanding of the difference between waking and dreaming to divide knowledge into categories. That which is real is based on perceptual observation compared to that which is not. This approach laid the groundwork that would lead to the experimental methodology that would lead to science. This is the question that Descartes was trying to address when he developed his scientific method, and moved the Western world into the structured cognitive analysis and awareness of waking reality that we rely on today.

What we accept as science today evolved out of the Cartesian tradition of analytical philosophy. In the West that approach began with Descartes' study of dreams. Yet dreams are also intrinsically part of other philosophical perspectives, including Eastern philosophy, as well as post-modern or post-structuralist philosophy. These traditions are prone to addressing the aesthetic and alternative (to waking) cognitive characteristics of the dream state. According to legend, Zhuangzi, the great Taoist philosopher, fell asleep one day and dreamed that he was a butterfly. When he woke up, he did not know whether he was a man who in reality dreamed that he was a butterfly or whether he was a butterfly, dreaming that he was a man (2). In Eastern philosophical tradition, this story has had a similar impact to that which Plato introduced in the Western tradition in his allegory of the Cave. This question of dreaming butterfly versus dreaming man addresses an aspect of the human condition that probably dates back to our specie's first cognitive awareness.

Descartes was one of the founders of Enlightenment science, the father of the mathematical science of geometry, and among philosophers, the great dreamer. He was the one who focused most on how dreaming affects our conception of waking reality. He proposed three major philosophical arguments based on the cognitive experience of dreaming:

- How does the veridical experience of external reality (truth) differ from the apparent truth of dreaming?
- What is the relationship between our physical body and our mental conceptualization of reality (dream/mind)?
- In studying the role of dreaming in memory, does memory in dreaming provide the tool that can be used to differentiate waking from dreaming?

These questions still have no concrete answers. They continue to be the focus for academic conferences and philosophical treatises. These questions remain at the basis of any discussion of dream philosophy and science likely to take place today.

As the "mysterian" philosopher Colin McGinn has famously noted: "In fact, when scientists ... try their hand at philosophy – usually after they have retired – the results are often inept, risibly so" (3). Accepting this judgment as probably true, this chapter on Dream Philosophy is written from a perspective outside and beyond the formal discourses of philosophy, by a currently unretired dream scientist and armchair admirer of Rene Descartes – the polymath–soldier who in the seventeenth century tore the world of the mind apart.

UNCONCEALING DREAMS

From the beginnings of philosophical method, dreams have proven to be an excellent philosophical topic in the discussion of "truth." Plato's parable of the Cave describes a situation, much like waking reality, in which we do not know whether to trust the evidence of our senses as to what is real or what may be reflected illusion. Descartes took this philosophical dilemma a step further in exploring the apparent reality of dreams:

> And finally, taking into account the fact that the same thoughts we have when we are awake can also come to us when asleep, without any of the later thoughts being true, I resolved to pretend that everything that had ever entered my mind was no more true than the illusions of my dreams .... For how does one know that the thoughts that come to us in our dreams are more false than others, given that they are no less vivid or expressed ... for our dreams ... represent to us various objects the same way as our exterior senses do ... what truth there is in them (our thoughts) ought infallibly to be found in those we have when awake rather than those we have in our dreams. (Descartes, 1641) (4)

The skeptical argument that one cannot differentiate dreaming from waking experience still resonates among philosophers today (5).

Descartes approached this question with what we now call the "scientific method," an approach that had been revealed to him in a series of dreams. He recorded these dreams in the year 1619, after returning to France from a battle in what is now called the Hundred Years War, in which he had been fighting as a mercenary soldier. In the first of these dreams, Descartes described being blown and twirled by the wind into a college churchyard where, still staggering, he was offered a melon (pomegranate) by a stranger who seemed to know him (as in the initial quote for this chapter). He woke from this dream in pain, in fear that "an evil genius" was bent on seducing him. The second dream had little content, but was filled with a "sharp and piercing noise" which he took to be a clap of thunder. On waking, he reports that his room was filled with fiery sparks. The third and final dream was much longer. It included two books, a dictionary, and an anthology of poems open to a page that included the statement: "*Est et non.*" Interpreting this dream while still in his dream, Descartes judged that the two books signified the union of philosophy and wisdom, the sparks being the seeds of wisdom of divine inspiration:

> By the poem, "*Est et non,*" which is the "yes" and the "no" of Pythagoras, he understood truth and error in our human knowledge and the profane sciences. Finding that he succeeded in fitting in all these things well to his satisfaction, he was so bold as to feel it was assured that it was the spirit of Truth that by his dream had deigned to open before him the treasures of all the sciences. *(Baillet, 1691) (6)*

Descartes, waking, interpreted this series of dreams to mean that he was the person destined to reform knowledge and unify the sciences, that the search for truth would define his career, and that the thoughts that he had been having in the previous month about a unifying system for analyzing knowledge and experiment could become a new method for attempting to find the truth. "I begin to understand the foundations of a wonderful discovery ... all the sciences were connected by a chain; no one can be completely grasped without taking the whole encyclopedia at once" (7).

His contemporaries and later philosophers were apologetic and even embarrassed that the father of analytical geometry and the scientific method had derived his realizations in part from dream, and then even had the audacity to write the dreams down. Auguste Comte commented: "Descartes had a cerebral episode," and "the passage in which he relates how his brain was over-stimulated as in a fit state for visions ... shows great weakness" (8). Christiaan Huygens tried to be kind: "He tired himself to such an extent that his brain became overheated and he fell into a kind of rapture which worked upon his already exhausted spirit that it became predisposed to the reception of dreams and visions." For the general public, these dreams served to bring Descartes into the popular culture. In that era, the dream of the melon by the great philosopher was considered quite ribald (9). That itself is an interesting comment as to how sexual mores, masturbatory practices (somehow involving vegetables such as melons), and our world have changed in the intervening centuries since Descartes.

Descartes utilized these dreams to develop his scientific method based on the following precepts: (i) never accept anything is true that is not known to be so; (ii) divide the problem into as many component parts as possible; (iii) build from the smallest of what is known; and (iv) keep complete records for review. Descartes' first philosophical use of this method can be found in his attempts to differentiate dreaming from waking reality (10).

During Descartes' era, "truth" was defined by the Christian Church. As a contemporary of Galileo, Descartes put off publishing his work while Galileo was under papal edict, and threat of death, unless he refuted the astronomical evidence that Earth was not the center of the universe. Descartes eventually obtained the support of clerics within the Church who became convinced that his analytical approach to science offered a new way to scientifically prove the existence of God. It was a younger philosopher who had once declared, "Some years ago I was struck by the large number of falsehoods that I had accepted as true in my childhood, and by the whole edifice that I had subsequently based on them. I realized that it was necessary, once in the course of my life, to demolish everything completely and start again right from the foundations if I wanted to establish anything at all in the sciences that was stable and likely to last" (11). It was the older philosopher who hid his radical thoughts inside Church dogma and the belief of his supporters that his scientific method could be used to demonstrate the existence of God; it was the older philosopher who was published and who became famous. He had to become the consummate philosopher in order to survive, willing to confront and confound the powerful Roman Catholic Church by redirecting and arguing the opposite side of his own arguments. He convinced his friend, the Jesuit Cardinal de Bérulle, that in the scientific details of the natural world evidence could be found to prove the existence of God, by using the logic of the new sciences. From this Catholic perspective the mind is that by which we know; all else is body, which is what is known. While Galileo fretted under house arrest, Descartes obtained the support of the Roman Catholic Church through the Jesuits. He then went off to Sweden to join Queen Christina's court. Descartes the philosopher proposed for his own tombstone a very concrete and politically appropriate epitaph: "He who hid well, lived well" (Ovid's *Tristia*: "*Bene qui latuit bene vixit*").

While Descartes stayed out of prison and led a moderately long and entertaining life, dancing and philosophizing in the court of Queen Christina, Western religions co-opted his attempts at deriving a philosophy of truth. Descartes' scientific method continued to be utilized both in side and outside the Church. However, within both the Church and society, the search for truth became a search for the absolute truth, to be found in the fundamentals of scripture. This early fundamentalism was acted out in a series of European wars and revolutions occurring in concert with the Protestant Enlightenment.

Ultimately, the cosmological evidence for an extended and non-human-centered universe, and Darwin's elegant theory of evolution, with its archaeological and geological evidence that our species has developed over extended rather than historic time, proved difficult to incorporate into fundamentalist Christian concepts of absolute truth based on a strict reading of the Bible. In a society defined by religion and an absolute concept of "truth," Enlightenment scientists incorporated religious concepts into their methodology. Science was based on equations that had mathematical precision. The Newtonian science of the era indicated that the relationship of every particle, once fully described, would explain the structure and interactions of matter – and perhaps even of mind (12).

Much later, as a historian of ideas, Heidegger argued that it was at this point – when truth became an absolute – that the Western world's approach to truth became illogical. Heidegger presented this argument based on his retranslation of Plato's allegory of the Cave and his redefinition of the Greek term (*aletheia* – unforgetting) that Plato had used as his term connoting "truth." Heidegger argued that for Plato, the essence of truth was "unconcealment" rather than absolute belief. Descartes' scientific method offered a tool that could be used to pull back layers of concealment to reveal a relative truth hidden underneath. He argued that such a perspective of truth is necessary if we are to function in this world, "… not dreaming, not falling into outlandish plans and unrealizable wishes, but just pursuing the everyday and contributing to its endurance" (13). Of course, Heidegger developed this approach to truth well after the development of relativity physics, quantum mechanics, and the realization in science that even the supposedly mathematically solid proofs of physics were statistically probabilistic and relative.

Today, there is little question that our perceptual experience of the external world, which seems so real to each of us, is but one of many of the possible coded subjective representations of external reality. What we perceive is always *projected* sense data. We live in a world of naïve realism where our consciousness acquaints us with its own contents, a play of images upon a mental screen (14).

Descartes developed a method that could be used in the attempt to differentiate dreaming from reality, but any such approach to "truth" remains exceedingly difficult. Today's philosophers tend to view truth in context. "The truth we are dealing with here is not 'objective' truth, but the self-relating truth about one's own subjective position; as such it is an engaged truth, measured not by its factual accuracy, but by the way it affects the subjective position of enunciation" (15). From this perspective, truth depends on the needs of the individual or the state. For some philosophers, the only requirement for establishing truth is that it be useful in helping to describe the structure and function of reality or the unity and coherence of illusion and dreams (16). For the modern philosopher who views truth as a matter of context, it is quite obvious that it was the man dreaming of the butterfly rather than the butterfly dreaming that it was a man. Men write books, butterflies do not (17).

Scientists continue to peel back the layers of the onion that conceal the essential truth of our reality. Religions, at least as powerful as science in defining the structure of our modern world, continue to emphasize time-less essential truths. These truths vary from religion to religion, and often result in conflicts among their adherents. In such a world, context is what matters. In such a world, dreams can be truths.

THE DREAM AS A WINDOW INTO BODY AND MIND

In applying his scientific method to differentiate waking reality from dreaming, Descartes created a structural paradox that still affects the sci-entific study of dreaming. Descartes postulated that there were two sorts of substances: mental and material. Within each individual, there is a union of material substance (body) and mental substance (mind). Body is composed of matter that has an external nature that can be divided, spatially extended, and described by logic. If a state can be scientifically studied, measured, classified, or logically proven to exist, it is not what Descartes defines as a dream. For Descartes, dreams are phenomena of mind (18).

During Descartes' time, this differential between body and mind was generally defined in religious terms, in which body was of this world and mind was of God. Conceptually, discussions of differences between spiritual mind and material body are to be found in the first written reli-gious texts.

> Jesus said, "If the flesh came into being because of spirit, it is a marvel, but if spirit came into being because of the body, it is a marvel of marvels. Yet I marvel how great wealth has come to dwell in this poverty." *(The Gospel of Thomas) (19)*

In this quote, attributed to Jesus as ancient skeptic, there is the sugges-tion that mind (spirit) is all that is not flesh. From some religious points of view, the concept of the separation of body and mind has been inter-preted to mean that all that is not flesh comes from outside the individ-ual. In other words, mind is God or god-like, and flesh is human.

Historically, the meaning of "mind," like the meaning of "melon," has changed as each generation has applied its unique cultural, social, reli-gious, and logical mores to defining the concept. There are periods of per-vasive cultural and religious conventions in which the definition of mind is so self-evident as to not require discussion. Periodically, studies into the relationship between body and mind have developed into new phil-osophical traditions. And between generations, the argument that body is radically different from mind would resurface as perspectives of body and mind changed, based on changing cultural mores and technology. The variability in what is meant by "mind" increases with each generation as older traditions persist in coexistence with newer perspectives. Today, mind as a philosophical concept has many meanings. A tradition has developed of avoiding conflicting definitions, presuming that what one individual is referring to when mentioning the concept of "mind" is the same as for all others. Philosophers faced with conceptual confusion have sometimes resorted to a definition of mind based on customary usage (20). Today, it is not unusual for authors to include "mind" in the title of their book and then never define or specifically address its meaning. Like truth, mind is defined by the context in which it is being discussed.

There are still many who accept Cartesian and Christian separation of body from mind. However, most scientists, philosophers, and theologians have moved beyond this spiritual/religious perception of Cartesian dualism. Outside religion, the body/mind differential has been structurally incorporated into epistemologies. Astronomy, subject to scientific method and experimental proof, is a science. Astrology, while also utilizing numerical analysis, is not subject to the constraints of logical proof; therefore, it is not a science. Chemistry is science; alchemy is not. These Cartesian borders are less distinct for other fields. Sociology is subject to the constraints of experimental science. Political science is not, but does that make it "mental" and an aspect of mind? Medicine, whenever possible, is based on scientific method, yet for diseases of the mind diagnosed and treated by psychiatry, this is not necessarily so. Philosophy once included the sciences. It was only when logical, empirical evidence became available, based on the Cartesian differential, that these new sciences were split off from the study of mind to become their own fields of body-based knowledge.

Since Descartes, the philosophy of mind has focused on whether or not a radical difference exists between body and mind. There is an assumed clear and basic difference between Cartesian dualists who think that mind and body are fundamentally different, and monists who think that mind and body are the same kind of phenomenon. Dualistic philosophers of mind are further categorically classified into "substance dualists" who think that body and mind are two different kinds of substances, and "property dualists" who think that "mental" and "physical" name two kinds of "properties" based on the same substance (e.g., the human brain). Among modern philosophers, Thomas Nagel and Colin McGinn are considered property dualists, while theologians who persist in their belief in "soul" can be viewed as substance dualists. Monists are often divided into "idealists" who think that everything is ultimately mental and "materialists" who think that everything is ultimately material or physical (21).

Among professionals in philosophy, psychology, neurobiology, and cognitive science, almost everyone in the past fifty years has become a materialist, believing it self-evident that mind must equal brain. The many materialist philosophers can be divided into various schools.

Behaviorism postulates that mental states are patterns of behavior and dispositions to behavior. Behaviors are the aspects of body that lead to cognitive processes (mind). While behaviorism was once popular, few philosophers and/or psychologists in today's world are fully behaviorists. Most take it for granted that mental events lead to physical effects (behaviors) rather than *vice versa*. We see a commercial and then we go shopping. We consider that this happens because our beliefs and desires (sometimes commercially induced) lead to the physical process, rather than the process somehow inciting our desire. While behaviors can affect states of mind, most now accept that it is mind that causes behavior.

Physicalism or identity theory postulates that mental states are equivalent to brain states. For the identity theorist who believes that body equals mind, mental processes are identical to neural processes. This mind–brain identity theory, sometimes called "monism," has, in the past fifty years become the most influential theory of mind. From the unitary perspective of monism, mind is defined as the set of cognitive processes based on brain activity (22). There is a plethora of unified monistic theories of mind–brain (23). As a result of the rapid eye movement sleep (REMS)-equals-dreaming debacle (discussed in detail in the Preface of this book), it has become more difficult to accept without question the various claims for neural correlates of mind. Yet many scientists continue to believe that mind is irreducibly a biological phenomenon, and that it is only the limitations of our current research capabilities that keep us from discovering the specific neuron, cellular process, or brain slice that functions as the biological basis for particular states of mind (24). Perhaps the mind is hidden in the frontal cortex, the hippocampus, the limbic system, or as Descartes proposed, in the tiny melon-shaped pineal gland.

The limitations of behavioral and identity materialist theories have led to the modern theory of functionalism, which combines features of both these belief systems. Functionalism suggests that mental states are physical not because of their physical characteristics, but rather because of their functional and causal relationships. To a functionalist, a state such as consciousness does not necessarily require a brain; it requires a system with capacities that, like the brain, can result in consciousness. According to Dennett, perhaps the most popular proponent of functionalism, mental states are externally evident and within the capacity of artificial intelligence (AI) systems. Based on his computational theory of consciousness, he has proposed that the mind is currently available for rigorous, perhaps definitive scientific study (25).

Since dualism seems unscientific, it is unacceptable to most philosophers and scientists. Physicalism and behaviorism are probably incorrect based on current empirical evidence and useless in attempting to explain

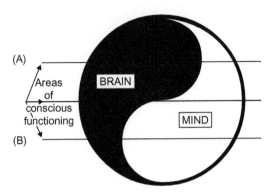

FIGURE 3.1 The Yin (mind)–Yang (body) classification diagram for conscious process. Each cognitive state or experience includes a component of both body and mind (note transecting lines). Primarily body-based states (e.g., skiing or playing a musical instrument) (A) also include a component that is mind that contributes to the optimal performance of that state. Primarily mind-based states (e.g., creativity or dreaming) (B) have a larger component that is mind. However, each is structured upon a component that is body (brain) based.

mental states (26). Functionalism awaits supporting evidence. In the general public, most individuals conceive of themselves as having both spiritual and physical characteristics that are not necessarily connected. They are at least relative dualists. I find myself in this category. Most systems of knowledge have characteristics that are both mind and brain (27) (Figure 3.1).

There is an obvious physicality to the state of sleep in which dreams occur. Dreams have physical, brain-based correlates such as imagery, emotion, and memories, but the correlate is partial and not nearly as simple as REMS equals dreaming (28). Although some modern scientists and philosophers argue otherwise, major aspects of dreaming still cannot be scientifically studied, measured, classified, or logically proven to exist. Dreams are not what Descartes defined as body or material substance. Much of the cognitive state that we call dreaming, including the meaning and function, must still be considered to be phenomena of the mind.

I find the property dualism espoused by McGinn to be particularly attractive. Writings in philosophy are often dryly and epistemologically overly intellectual. McGinn has the writing skill and the intellectual flexibility to reach beyond those limitations. He argues that our ability to perceive "mind" is inherently limited by the cognitive characteristics of our species: "It is rather that mental concepts are intuitively such that no physical concepts could characterize the essential nature of the mental process denoted" (29). He argues that we suffer from "cognitive closure" with respect to the mind–body problem, that "mind" is on the other side of a border, which scientific technical capacities built on our operative logical cognition are unable to cross and unable to see. This perspective, while aesthetically and logically attractive, if coupled with Descartes' Cartesian differential and taken to the extreme in which dreams are viewed only as aspects of mind, can lead to the conviction that dreams cannot be scientifically studied.

John Searle argues, perhaps correctly, that we have inherited an obsolete Cartesian classification system, and that this system of philosophical classification is interfering with modern approaches attempting to clarify the problem of consciousness (30) (to be discussed more fully in Chapter 5). Maurice Merleau-Ponty goes even further to suggest that, philosophically, no difference exists any longer between the many Cartesian perspectives, "because dualism has been pushed so far that the opposites, no longer in competition, are at rest the one against the other, coextensive with one another" (31). However, looking at the current status of the mind – body debate, the philosophical debate addressing mind and brain appears to be in a much better and less destructive place than the debate surrounding Descartes' other basic question – the essence of truth – addressed at the start of this chapter. This book is being written during another presidential election in America (Romney vs. Obama), when the airways are full of cleverly developed propaganda based on the newest philosophical concept of absolute truth in context. We have, at least, the option of turning off the context of that world to spend a bit of our psychic energy considering dreams as having aspects of both body and mind.

MEMORY, DREAMING, AND FRONTAL LOBOTOMIES

Descartes concluded that in perceptual isolation, man was at the center of his own universe.

> But I know now with certainty that I am, and at the same time it could happen that all these images – and generally, everything that pertains to the nature of the body are nothing but dreams ... But what then am I? A thing that thinks. What is that? A thing that doubts, understands, affirms, denies, wills, refuses, and which also imagines and senses. *(Descartes, 1641, p. 68) (32)*

Working from that original description of "self," Descartes went on to differentiate his dreaming world from his reality without having to rely on what he considered to be the sometimes undependable evidence of his senses:

> I should no longer fear lest those things that are daily shown me by my senses are false; rather the hyperbolic doubts of the last few days ought to be rejected as worthy of derision – especially the principal doubt regarding sleep, which I did not distinguish from being awake. For I now notice that a very great difference exists between these two; dreams are never joined with all the other actions of life by the memory, as is the case with those actions that occur when one is awake. *(Descartes, 1641, p. 100) (33)*

Descartes' dream-based scientific method for assessing empirical truth lies at the basis of modern science. His insights into the differential between dreaming and waking reality have led to an entire field, the philosophy of mind. His insights into memory have received less attention than these other dream-associated philosophical musings. Yet today Descartes' emphasis on memory seems both remarkable and prescient. The philosophical topics of truth and mind have been debated *ad nauseum* and have become a bit worn. And today, the scientific study of memory is proving to be among the most useful approaches to the study of cognitive states – including dreaming.

In general usage, the term "memory" is used as a generic concept for information storage. Dreaming incorporates internal memory processes that are not concerned with the processing of externally derived perceptions and stimuli. Memory assumes the priority role and function in such internal cognitive processes that build on stored experience. In everyday life such memories are incorporated into dreams and daydreams, plans, evaluations, and reasoning that take place covertly. These memory systems are difficult to study since most of these internal cognitive experiences never result in observable actions or externally identifiable responses (34).

Different forms of memory are involved at different stages of dreaming. Real memories of experienced events are incorporated and represented within the dream, contributing to dream content. Memory processes function within the dream to organize, sequence, and monitor the dream narrative. The dreamer upon awakening, in order for the dream to be recalled, must remember the dream. The intensity and persistence of what is recalled affect the ways in which the dream is used in the dreamer's waking activities. At each of these levels, different forms of memory are utilized. Experimentally, different variables affect each of these levels in which memories compose, integrate, and structure the dream remembered in waking (35).

As is the case with most of dream science, memory studies have focused almost exclusively on the relationship between REMS and memory processing. Not long after the discovery of REMS, connections were discovered between sleep and intelligence, learning and memory processes (36). Only a few researchers persisted with these lines of research. The majority were pulled into the study of dream content, psychoanalytic constructs, and neuroconsciousness. The exception has been in one area of neuroconsciousness theory: the Crick–Mitchison reverse learning theory postulates that dreaming/REMS is involved in eliminating spurious memories during the dream state. Based on this theory, dreams are meaningless constructs of spurious sense and memory data that need to be excluded from the system in order for it to function properly. In relation to information processing and computing, this theory has a certain elegance of presentation. However, like other neuroconsciousness theories, this theory is clouded by its requirement that REMS equals dreaming (37). There is little actual evidence or theoretical support for the suggestion that dreaming functions in reverse learning.

Researchers are now re-examining potential functions for sleep and dreaming in learning and memory. A beneficial role for sleep in learning and long-term memory consolidation is of the few functions for sleep that is supported by experimental evidence (38). Memory is composed of different subsystems that utilize different neural structures and function independently of each other (Table 3.1). Long-term memory (consolidated memory) can be divided into two major systems, declarative and non-declarative. When using declarative memory the subject is consciously aware of easily accessible stored information. Declarative memory can be subdivided into semantic memory (our factual knowledge of the world) and episodic memory, which includes the contextual location of that memory in time and space. Non-declarative memories are learned without conscious awareness. They are not as easily accessed or as easily described. Subtypes of non-declarative memory include procedural memory (the learning of skills and habits), priming, classic conditioning, and emotional memory (39).

Most studies of sleep and memory consolidation experimentally approach this question by depriving subjects of normal sleep. Sleep deprivation leads to sleepiness during waking and difficulty with focus and concentration. Because of this resultant daytime sleepiness, sleep deprivation negatively affects many of the waking tasks used to access memory systems. These are experimentally powerful effects that are difficult to control. They affect and alter the results of any research work that uses sleep deprivation as an experimental approach (40). Stage-specific sleep deprivation studies suggest that different stages of sleep may be differentially beneficial for specific subsystems of memory processing. It is likely that different kinds of memories are consolidated in different stages of sleep (41). The non-REM stages of deep sleep and stage 2 sleep are apparently involved in the consolidation of declarative learning (42). The evidence supporting a role for REMS in memory consolidation is less clear.

TABLE 3.1 A Classification System for Consolidated Memory

Long-term (Consolidated Memory)	
Declarative	**Non-Declarative**
Semantic	Procedural
Episodic	Priming
	Classic conditioning
	Emotional

The best evidence is for a supporting role for REMS is in the consolidation-based enhancement of non-declarative memories (43). Potentially, REMS serves a role in emotionally charged memory processing, for example, in the nightmares of post-traumatic stress disorder (PTSD). This perceived connection has contributed to recent proposals for the functioning of dreaming in emotional processing (44).

It is even less clear that the process of dreaming (decoupled from its dissociable correlate of REMS) has a role in learning and the consolidation of memory. Viewed independently of sleep stage association, dream content can reflect recent learning, and may be helpful in creating new approaches to problem solving (45). Recurrent negative dream experiences (nightmares) are the most common symptom of PTSD. Such distressing dreams are likely to reflect a breakdown in the emotional processing system utilized to process major physical and psychological trauma (46). The evidence that dreaming contributes to the process of memory is quite limited, and what evidence there is relates primarily to the role of nightmares as a marker for dysfunction in the cognitive systems utilized in the processing of emotional memories.

THE SAME MEMORY SYSTEMS ARE USED IN DREAMING AND WAKING

Perhaps Descartes was right, and the study of memory is the key to differentiating the dream reality of sleep from waking experience. But there is strong evidence that the same memory systems are utilized in both dreaming and waking. This conclusion is based on several lines of evidence.

(i) Memory processes active in the processing of external waking are also active during the processing of memory in dreaming

Memory-based processes active during dreaming include visual imagery, automatic and attentional cognitive operations, emotional competent memory systems, and self-reference. Waking imagery, like dreaming, is a system of cognitive processing that is dissociated from perceptual input. Concepts developed from the study of waking imagery (e.g., the ability to develop "mental images" of objects without actual perceptual input) are also characteristic of dreaming memories (47). Mental imagery differs from perceptual images in several basic ways: it is slower and more controlled without the stimulus-based attention shifting that is characteristic of external perception; imaged objects fade quickly while external objects persist as long as they are present; the images in imagery are limited by the information encoded into memory; and we have control over objects in images while having little or no control over perceptual images based on the external world (48). Like dreaming, imagery is primarily visual and replete with multiple memory associations (49). Because imagery does not include perceptual input, it cannot be experimentally studied with the same techniques used to study perception-based memory. However, visual aspects of both imagery and dreams can be studied using similar approaches at multiple levels of processing, including image generation, image inspection, image maintenance, and image transformation.

The cognitive architecture required for non-perceptual image generation requires these interacting memory processes. Techniques have been developed for experimentally evaluating the transfer of imagery into conscious awareness. In imagery, the visual representation of an object is achieved with either a global shape or part of a shape that is integrated with associative memory into the representation of a particular object. Multipart images can be integrated using object and category spatial relationships to form an image. These spatial relationships can be described as the operative series depicted in Table 1.1 (see Chapter 1) (50). This descriptive paradigm of the cognitive process of imagery can be applied to the operative image processing occurring during dreaming. These same techniques can be used to study the organization and expression of non-perceptual memory in dreams (51). Dreams integrate and use these internal constructions and processes of memory; the same systems are used in processing our sensory perceptions of external reality.

Dream images, although they may appear internally to be images from the real world, are as thin as thought. The thought space of such imagery is very different from perceptions of real space (Chapter 9) (52). The reduction of imaginative "seeing" to a brain-based operative cascade parodies the visual experimentation of the impressionist and post-impressionist artists. It was Picasso who stated that, "it would be interesting to preserve photographically … the metamorphosis of a picture. Possibly one might then discover the path followed by the brain in materializing a dream" (53).

(ii) Interference, retention interval, and salience, basic concepts used to characterize memory, were developed during the study of dream recall

Dream recall declines rapidly with increasing time since the dream, dropping with increasing speed after waking (54). If you wake gradually rather than quickly, your ability to recall your dreaming declines (55). Koukkou and Lehmann incorporated this memory trace concept into their "arousal–retrieval" model. They demonstrated that short-term dream recall declines rapidly based on the time since the dream occurred and the time since arousal from the sleep memory-based trace of the dream. The dream trace may be available for only a short time (seconds or minutes) for incorporation into the long-term memory store accessible during waking (56). This decline in dream recall with time after waking has been applied to the study of imagery, and as with dreaming, imaged objects fade more quickly after the cognitive experience than do perceptually based images. Distraction and rapid cognitive-process change occurring after waking negatively affect dream recall (57). Interference also affects imagery and perceptual memory consolidation. Conceptually, the cognitive processes of both interference and retention interval developed from the study of dreaming have been reapplied to the study of waking memory (58). This short-term memory trace, sometimes referred to as "working" memory, is described best in the processes of dream and imagery recall. It is likely that working memory utilizes a representational cognitive processing system to construct images and symbols that exist virtually as visual constructs in our brains (see Chapter 9).

Salience accounts in part for the relationship between dream recall and content. Dream recall is affected by dream intensity, with the most intense or "transforming" dream more likely to be remembered. Dream salience – the greater novelty, bizarreness, affectiveness, or intensity of an experience – increases the potential for recall of a particular dream. Nightmares, a primary example of dream salience, are often remembered in great detail after waking. This finding of increased recall with increased salience is also found in perception-based waking memory and in imagery (59).

(iii) Dream recall is preserved after most major neurosurgery and even after extensive neuroanatomical damage to the brain that eliminates REMS

Patients who have lost as much as half of their brain content to surgery still dream. It is only in cases of severe bi-basilar frontal central nervous system (CNS) damage (injuries similar to those seen secondary to the psychosurgery called "frontal lobotomy") that a loss of dreaming is observed (60). Even in these cases, dreaming often reoccurs with time (61). Some forms of brain injury, particularly those affecting visual systems involved in waking memories, can have the same effects on visual aspects of dreaming (62). This finding is inconsistent and varies based on the area of injury. Some patients who demonstrate major CNS-based visual defects do not have changes in their dream images (63). Even in cases where extensive neuroanatomic damage results in physical and psychological disability, dreaming is preserved. Dreaming must be important.

It appears that Descartes was prescient in focusing on memory in dreams, but it also appears that memory, rather than being the marker demonstrating the differences between waking and dreaming realities, is one of the key indicators demonstrating that dreaming is in part brain based. Memory is primarily a brain-based process. Dreaming may not be involved in the actual processing of memory. However, the evidence is quite strong that dreaming uses the same memory processing and systems utilized in waking perceptual memory. Most researchers in the field of memory and sleep believe that the memory processes used internally in dreaming are the same systems used in waking and externally derived memories. The component of dreaming that is memory is most likely to be brain based.

DESCARTES' MANY ERRORS?

Some modern neuroscientists, such as Damasio, have staked their reputation on their belief that Descartes was wrong (64). However, there is good evidence that in his perspectives of dreaming, Descartes was correct. Beyond brain-based components of memory and visual imagery, dreams include aspects of mind-based process and content. Descartes' cohorts were probably right in their belief that his dream melon was pornographically suspect; but they were wrong as well. It was a dream melon, and the possible associations leading from that melon into Descartes' internal mental constructs, just like the insights available from his philosophy, seem endless. There are aspects of mind such as memory that can be scientifically studied. There are aspects that cannot. Systems of memory give us a logical time-based framework through which to comprehend our external world. These systems are present in dreams. But dreaming as an aspect of mind includes cognitive processes that we cannot see even with modern technical extensions of our perceptual expertise. Dreams are windows looking in from waking consciousness into an otherwise non-conscious internal world of our minds.

Philosophically and scientifically, dreams are an ideal model for addressing questions of truth (unconcealment), and for exploring the border between body and mind. Descartes proposed that studying dreaming would lead to a better understanding of the cognitive processes that we use to understand external waking reality. But dreaming offers much more. The components of dreaming that are brain-based provide routes by which scientific method can approach and empirically study non-conscious mental functioning; exploratory routes that can be utilized to approach that modern-day philosopher's stone – the origin of consciousness.

Notes

1. Baillet, A. (trans.) (1691) Descartes' three-fold dream, in *La Vie de Monsieur Descartes*. Paris: Chez d'Horthemels, pp. 81–86.
2. Palmer, M. (1996) *The Book of Chuang Tzu*. London: Penguin Books.
3. McGinn, C. (2002) *The Making of a Philosopher: My Journey Through Twentieth-Century Philosophy*. New York: HarperCollins, p. 200.
4. Descartes, R. (1641) Objections against the meditations and replies, in *Great Books of the Western World: Bacon, Descartes and Spinoza* (1993), ed. M. J. Adler.

Chicago, IL: Encyclopaedia Britannica, pp. 17, 20–21. Descartes, like many older philosophers, is now better known for his contributions to what would be future philosophical approaches than for his original work. Modern scientists and philosophers (e.g., Damasio and Dennett) are quick to identify errors and authoritarian concrete concepts with Descartes. As originally conceived, the original works are quite cognitively flexible.

5. Blumenfield, D. and Blumenfield, J. B. (1978) Can I know that I am not dreaming?, in *Descartes: Critical and Interpretive Essays*, ed. M. Hooker. Baltimore, MD: Johns Hopkins University Press, pp. 235–255.
6. Baillet, A. (trans.) (1691) Descartes' three-fold dream.
7. Clark, D. (2006) *Descartes: A Biography*. Cambridge: Cambridge University Press, pp. 58–59.
8. Harman, W. and Rheingold, H. (1984) *Higher Creativity: Liberating the Unconscious for Breakthrough Insights*. New York: Tarcher/Putman, p. 77.
9. Vrooman, J. (1970) *Rene Descartes – A Biography*. New York: Putnam Press.
10. Descartes, R. (1941) *Meditations on First Philosophy*, trans. D. A. Cress (1980). Indianapolis, IN: Hackett.
11. Descartes, R. (1637) *Discourse on Method*, trans. D. A. Cress (1980). Indianapolis, IN: Hackett.
12. Bechler, Z. (1991) *Newton's Physics and the Conceptual Structure of the Scientific Revolution*. New York: Springer.
13. Heidegger, M. (1988) *The Essence of Truth*, trans. T. Sadler (2002). New York: Continuum, pp. 35, 54–55.
14. McGinn, C. (1991) *The Problem of Consciousness*. Oxford: Blackwell.
15. Zizek, S. (2012) *Living in the End Times*. New York: Verso.
16. Searle, J. (1992) *The Rediscovery of the Mind*. Cambridge, MA: MIT Press.
17. Tallis, R. (2012) Zhuangzi and that bloody butterfly, In *Defense of Wonder – And Other Philosophical Reflections*. Durham: Acumen, p. 30. Tallis really has no chance against the butterfly.
18. Kim, J. (1998) *Philosophy] of Mind*. Boulder, CO: Westview Press, pp. 47–52.
19. *The Gospel of Thomas: The Nag Hammadi Text* [29] (*c*. 100 AD).
20. Churchland, P. (1986) *Neurophilosophy – Towards a Unified Science of the Mind–Brain*. Cambridge, MA: MIT Press, p. 272.
21. Searle, J. (1997) *The Mystery of Consciousness*. New York: A New York Review Book, p. 135.
22. Place, U. (2004) *Identifying the Mind*, ed. G. Graham and E. Valentine. Oxford: Oxford University Press, pp. 58–59.
23. Churchland, P. (1986) *Neurophilosophy*, p. 4.
24. Searle, J. (1997) *The Mystery of Consciousness*, p. 195.
25. Dennett, D. (1997) An exchange with Daniel Dennett, in Searle, J. *The Mystery of Consciousness*. New York: A New York Review Book, pp. 115–131.
26. McGinn, C. (2002) *The Making of a Philosopher*, p. 178; Searle, J. (1997) *The Mystery of Consciousness*.
27. Pagel, J. F. (2008) *The Limits of Dream – A Scientific Exploration of the Mind/Brain Interface*. Oxford: Academic Press/Elsevier, pp. 189–194.
28. Pagel, J. F. (2011) REMS and dreaming – Historical perspectives, in *Rapid Eye Movement Sleep – Regulation and Function*, ed. B. N. Mallick, S. R. Pandi-Perumal, R. W. McCarley and A. R. Morrison. Cambridge: Cambridge University Press, pp. 1–14. the concept of relative dualism is discussed in detail in the last chapters of this book. Even clearly brain-based motor activities (e.g., running and skiing) include aspects that are apparently non-brain based and opaque to our technology.
29. McGinn, C. (1982) *The Character of Mind*. Oxford: Oxford University Press, p. 19.
30. Searle, J. (1984) *Minds, Brains and Science*. Cambridge, MA: Harvard University Press.
31. Merleau-Ponty, M. (1968) *The Visible and the Invisible*, ed. C. Lefort, trans. A. Lingis. Evanston, IL: Northwestern University Press, p. 54.

32. Descartes, R. (1641) Objections against the meditations and replies, p. 68.
33. Descartes, R. (1641) Objections against the meditations and replies, p. 100.
34. Cohen, G. (1996) *Memory in the Real World*. Hove: Psychology Press, p. 22.
35. Pagel, J. and Vann, B. (1997) *Cognitive Organization of Dream Mentation – Evidence for Correlation with Memory Processing Systems*. ASDA Abstracts; Cohen (1996) *Memory in the Real World*, p. 291.
36. Pagel, J. F., Pegram, V., Vaughn, S., Donaldson, P. and Bridgers, W. (1973) REM sleep and intelligence in mice. *Behavioral Biology* 9: 383–388. This first paper by the first author, published at 22 years of age, is among the first to present evidence for an association between sleep, memory consolidation, and intelligence.
37. Crick, F. and Mitchison, G. (1995) REM sleep and neural nets. *Behavioral Brain Research* 69: 147–155.
38. Walker, M. (2005) A refined model of sleep and the time course of memory formation. *Behavioral and Brain Sciences* 28: 51–104. When queried as to the evidence for a primary role for REMS in emotional memory, at the Association of Sleep Societies Meeting in Minneapolis (2011), Walker pointed out that, "It would not be the first time that I have been wrong."
39. Squire, L. (2004) Memory systems in the brain: A brief history and current perspective. *Neurobiology of Learning and Memory* 82: 171–177.
40. Siegel, J. (2005) The incredible shrinking sleep–learning connection. *Behavioral and Brain Science* 28: 82–83; Vertes, R. (2005) Sleep is for rest, waking consciousness is for learning and memory – of any kind. *Behavioral and Brain Science* 28: 86–87.
41. Stickgold, R., Hobson, J., Fosse, R. and Fosse, M. (2001) Sleep, learning and dreams: Off-line memory processing. *Science* 294: 1052–1057.
42. Smith, C. (2010) Sleep, memory and dreams, in *Dreaming and Nightmares*, ed. J. F. Pagel. Philadelphia, PA: W.B. Saunders, pp. 217–228. Stickgold, R., James, L. and Hobson, J. (2000) Visual discrimination learning requires sleep after training. *Nature Neuroscience* 3: 1237–1238.
43. Walker, M. (2005) A refined model of sleep and the time course of memory formation.
44. Levin, R., Fireman, G. and Nielsen, T. (2010) Disturbed dreaming, and emotional dysregulation, in *Dreaming and Nightmares – Sleep Medicine Clinics*, Vol. 5, ed. J. F. Pagel. Philadelphia, PA: W.B. Saunders, pp. 229–240.
45. Stickgold, R. (2003) Memory, cognition and dreams, in *Sleep and Brain Plasticity*, ed. P. Maguet, C. Smith and R. Stickgold. Oxford: Oxford University Press, pp. 17–39.
46. Levin, R. and Nielsen, T. (2009) Nightmares, bad dreams and emotional dysregulation: A review and new neurocognitive model for dreaming. *Current Directions in Psychological Science* 18: 84–88.
47. Pagel, J. (2008) Cohen, D. B. (1979) *Sleep and Dreaming: Origins, Nature and Function*. New York: Pergamon; Kosslyn, S. (1994) *Image and Brain – The Resolution of the Imagery Debate, A Bradford Book*. Cambridge, MA: MIT Press.
48. Kosslyn, S. (1994) *Image and Brain – The Resolution of the Imagery Debate*. Cambridge, MA: MIT Press.
49. Marks, D. (1990) On the relationship between imagery, body and mind, in *Imagery – Current Developments*, ed. P. Hampson, D. Marks and J. Richardson. New York: Routledge, pp. 1–36.
50. Kosslyn, S. (1994) *Image and Brain*.
51. Wolman, R. and Kozmova, M. (2006) Last night I had the strangest dream: Varieties of rational thought processes in dream reports. *Consciousness and Cognition* doi: 10.1016/j.concog2006.09.009; Kozmova, M. (2012) Dreamers as agents making strategizing efforts exemplify core aggregate of executive function in non-lucid dreaming. *International Journal of Dream Research* 5: 47–67.
52. States, B. (1997) *Seeing in the Dark: Reflection on Dreams and Dreaming*. New Haven, CT: Yale University Press, p. 97. Bert States made the crossover from the field of dramatic

arts to the study of dreaming, bringing an inquisitive and quick mind to the field that was unobstructed by current and past dogmas. He died much too soon.

53. Leger, F. (1938) *Fonctions de la Peinture*. Paris: Gothier.
54. Goodenough, D. R., Lewis, H. B., Shapiro, A., Jaret, L. and Sleser, I. (1965) Dream reporting following abrupt and gradual awakenings from different types of sleep. *Journal of Personality and Social Psychology* 2: 170–179.
55. Goodenough, D. R. (1991) Dream recall: History and current status of the field, in *The Mind in Sleep*, ed. S. J. Ellman and J. S. Antrobus. New York: John Wiley and Sons, pp. 143–171.
56. Koukkou, M. and Lehmann, D. (1983) Dreaming: The functional state-shift hypothesis, a neuropsychophysiological model. *British Journal of Psychiatry* 142: 221–231.
57. Kramer, M. (2007) *The Dream Experience: A Systemic Exploration*. New York: Routledge. Milt is old school. He has devoted his career to the attempt to bring structure to a sometimes undisciplined field.
58. States, B. O. (1993). Bizarreness in dreams and fictions, in *The Dream and the Text*, ed. C. S. Rupprecht. Albany, NY: SUNY Press, pp. 13–31. Goodenough, D. R. (1991) Dream recall; Kramer, M. (2007) *The Dream Experience*.
59. Kramer, M. (2007) *The Dream Experience*; Cohen G. (1996) *Memory in the Real World*; Kuiken, D. and Sikora, S. (1993) The impact of dreams on waking thoughts and feelings, in *The Functions of Dreaming*, ed. A. Moffitt, M. Kramer and R. Hoffman. Albany, NY: SUNY Press, pp. 419–476.
60. Kaplan-Solms, K. and Solms, M. (2000) *Clinical Studies in Neuro-Psychoanalysis: Introduction to an in Depth Neuropsychology*. London: Karnack Books, pp. 44–57.
61. Solms, M. (1997) *The Neurophysiology of Dreams: A Clinico-anatomical Study*. Mahwah, NJ: Lawrence Erlbaum.
62. Kerr, N. (1993) Mental imagery, dreams and perception, in *Dreaming as Cognition*, ed. C. Cavallero and D. Foulkes. Hemel Hempstead: Harvester Wheatsheaf, pp. 18–37.
63. Solms, M. and Turnbull, O. (2002) *The Brain and the Inner World*. New York: Other Press, pp. 208–211.
64. Damasio, A. (1994) *Descartes' Error: Emotion, Reason and the Human Brain*. New York: Avon Books.

Studying Dreaming and Consciousness

A toad the power mower caught,
Chewed and clipped a leg, with a hobbling hop has got
To the garden verge, and sanctuaried him
Under the cineraria leaves, in the shade
Of the ashen heartshaped leaves, in a dim,
Low, and final glade. **(Wilbur, "Death of a toad," 1950) (1)**

Consciousness is difficult to define and difficult to study. It includes
a spectrum of experiences that extend from states that are barely aware
to ones that are startlingly intense and remarkably vivid. Intellectually, it
can be frustratingly intangible. As noted in the *International Dictionary of
Psychology*:

> Consciousness: The having of perceptions, thoughts and feelings; awareness. The
> term is impossible to define except in terms that are unintelligible without a grasp of
> what consciousness means. Many fall into the trap of confusing consciousness with

self-consciousness – to be conscious it is only necessary to be aware of the external world. Consciousness is a fascinating but elusive phenomenon: it is impossible to specify what it is, what it does, or why it evolved. Nothing worth reading has been written about it (2).

Consciousness possesses the characteristics of an absolute metaphor: something known to exist which cannot be otherwise described (3). Consciousness may be best characterized as the subjective as opposed to the objective quality of experience. In general use, it is defined in multiple ways. For physicians, it is important to clinically define consciousness since the differential is sometimes between life and death.

In the medical clinic, consciousness is defined by a patient's arousability in response to stimuli – the individual's "awakeness." Since the term "consciousness" is definitionally diffuse, for the purpose of research and discussion the term is often restricted and used to refer to specific cognitive capacities, such as the ability to introspect or report one's mental status. Consciousness is sometimes considered synonymous with awareness. Being conscious is sometimes identified as the ability to focus attention or voluntarily control one's behavior. Sometimes the term "being conscious of something" has the same meaning as "to know about something" (4). A short survey of such attempts to address the definition of consciousness by some of our best and brightest philosophers and scientists is included in Table 4.1.

Philosophical and neuroscientific approaches to consciousness often limit themselves to characteristics of the waking experience that can be studied. There are those characteristics most often classified as components

TABLE 4.1   Example Explanations of Consciousness

"All consciousness is consciousness of something."	Kant (1781)
"Scientific points of view, according to which my existence is a moment of the world's, are always naïve and at the same time dishonest, because they take for granted, without explicitly mentioning, it, the other point of view, namely that of consciousness, through which from the outset a world forms itself round me and begins to exist for me."	Merleau-Ponty (1945, p. ix)
"A process in which information about multiple individual modalities of Sensation and perception is combined into a unified multidimensional Representation of the state of the system and its environment, and integrated with information about memories and the needs of the organism, generating emotional reactions and programs of behavior to adjust the organism to the environment ..."	Thatcher and John (1977, p. 294)

(Continued)

TABLE 4.1 (Continued)

In our naiveté, it seems now that conscious states are a single, unified natural kind of brain state, but we should consider the possibility that the brain is fitted with a battery of monitoring systems, with varying ranges of activity, and with varying degrees of efficiency, where consciousness may be but one amongst others, or where these systems cross-classify what we now think of conscious states. States we now group together as conscious states may no more constitute a natural kind than does say, dirt, or gems, or things-that-go-bump-in-the-night."	Churchland (1983, p. 92)
"… a matter not of arrival at a point but rather a matter of a representation exceeding some threshold of activation over the whole cortex or large parts thereof."	Dennett (1991, p. 166)
"We believe that it is hopeless to try to solve the problems of consciousness by general philosophical arguments: what is needed are new suggestions for experience that might throw light on these problems."	Crick (1994, p. 19)
"1. Not all the operations of the brain correspond to consciousness. 2. Consciousness involves some form of memory, probably a very short term one. 3. Consciousness is closely associated with attention."	Crick (1994, p. 15)
"… it is not possible to give a definition of 'consciousness' …"	Searle (1998)
"The consciousness of the mind is the combined result of the electrical activity of the brain."	Blaha (2000, p. 253)
"Consciousness is the crutch of cognitive neuroscience, perhaps the most widely used covert tool for the classification of mental phenomena. Yet this frequently used hypernym does not even have a definition. Is it a product, a process, or a thing? There is not even good agreement what the theory about consciousness would be like."	Buzsaki (2006, p. 360)

Sources: Kant, I. (1781) *Critique of Pure Reason*, trans. N. Kemp Smith (1929). London: Palgrave Macmillan.

Merleau-Ponty, M. (1945) *Phenomenology of Perception*, trans. C. Smith (1958). New York: Routledge Classics.

Thatcher, R. W. and John, E. R. (1977) *Foundations of Cognitive Processes*. Hillsdale, NJ: Erlbaum.

Churchland, P. S. (1983) Consciousness: The transmutation of a concept. *Pacific Philosophical Quarterly* 64: 80–93.

Dennett, D. (1991) *Consciousness Explained*. Boston, MA: Little, Brown & Co.

Crick, F. (1994) *The Astonishing Hypothesis: The Scientific Search for the Soul*. New York: Charles Scribner's Sons.

Searle, J. R. (1998) *The Rediscovery of the Mind*. Cambridge, MA: MIT Press.

Blaha, S. (2000) *Cosmos and Consciousness*. Auburn, NH: Pingree Hill.

Buzsaki, G. (2006) *Rhythms of the Brain*. Oxford: Oxford University Press.

of primary consciousness: the special sensory aspects of perception (auditory, visual, tactile, taste, olfactory), as well as the experience of pain and other body sensations (temperature and proprioception). Secondary consciousness includes those cognitive processes involved in integrating these perceptual and sensory inputs. Second-level cognitive processes include thoughts, emotions, the sense of self, learning, mental imagery, and orientation (5). Executive-level (tertiary) consciousness includes the cognitive functions involved in the integration of primary and secondary cognition. Executive-level functions include organization, planning, creativity, self-reflective awareness, execution of goal-directed behavior, flexibility of response to changing contingencies, task persistence despite distraction, and dream incorporation into waking behavior (6). Tertiary functions allow us to operate in a complex universe and society. Reflexive consciousness, required for dreaming, is the ability to self-consider the working of our brains and bodies. The thought capacities involved in reflexive consciousness include digression, revision, provisionality, uncertainty, free association, and free relapse into memory – the constant instability of a subject that knows itself to be altered by the act of thinking (7). Reflexive consciousness is a tertiary and executive central nervous system (CNS) function.

Consciousness is complicated. From a methodological standpoint, most theorists of consciousness restrict themselves to limited criteria such as focus of attention or control of behavior. Others focus on tests of primary or secondary aspects of consciousness. Theorists have developed potential tests for consciousness that can be used to test artificial intelligence systems and to determine the extent that animals have emerging aspects of consciousness. The most common of these tests are the Turing test, the mirror test, and the Chinese room. Debate persists as to the capacity of such tests for determining the presence or absence of consciousness (8). Consciousness is such a poorly defined state that the results of such testing reflect more on how the testing might apply to the definition of consciousness than on whether the animals or computer systems tested are actually displaying aspects of consciousness.

Animals demonstrate cognitive capacities for learning, remembering, problem solving, rule and concept formation, perception, and recognition (9). Such cognitive capacities are species specific, accomplished by some species much more easily than by others (10). Animals are observably able to process and respond in their behavior to perceptual information. This form of cognition is a form of primary consciousness. It is much less clear whether animals have the capacity for subjective experience (11). Self-awareness, based on the ability to recognize a mirror image as a representation of one's own body, is a capability that is clearly demonstrated by humans, chimpanzees, orangutans, and gorillas who have been trained to use communicative gestures with human trainers (12). These studies raise the question as to whether some animals may have a "theory of mind" – the ability to think about the mental experiences as well as the behaviors of others (13). The archaeologist Sophie de Beaune suggests that the non-hominin apes differ from humans in having a lack of analogical process, the process that humans use when faced with a new situation or problem of searching through their memories for an anterior experience that may apply to the new situation. This process requires referential knowledge, abstraction, and generalization (14). Its absence could account for the perceived lack of inventiveness present in the other great ape species (15).

Darwin, among others, was impressed that sleeping dogs sometimes move and vocalize in ways that suggest that they are dreaming. Early theorists of rapid eye movement sleep (REMS) equaling dreaming suggested that human eye movements in REMS might reflect the activity or experience about which the person is dreaming (e.g., the eyes move back and forth because the person is dreaming of a tennis match) (16). This perspective was extended by Damasio (1999) to the conclusion that animals experience self-referential dreaming: "… in dream sleep, during which consciousness returns in its odd way, emotional expressions are easily detectable in humans and animals" (17). Despite his contention, there is little evidence to support this supposition in humans. More than twenty percent of human sleep includes the repetitive conjugate eye movements of REMS, and dreams of tennis or ping-pong are a rarity. Among individuals with the diagnosis of REM behavior disorder who lose the motor blockade associated with REMS and apparently act out dream behaviors, it has been difficult to correlate those behaviors with dream content (18). If, however, Damasio is correct and sleep-associated activity reflects emotional expression, other animal species can be considered to demonstrate tertiary-level reflexive consciousness.

Alan Turing (1950) developed his "Turing test" in the attempt to determine whether or not computational systems had passed the threshold of becoming conscious. Turing originally described the test as follows:

> It is played with three people, a man (A), a woman (B), and an interrogator (C), who may be of either sex. The interrogator stays in a room apart from the other two. The object of the game is for the interrogator to determine which of the two is the man and which is the woman. The interrogator is allowed to question A and B, but not as to physical characteristics, and is not allowed to hear their voices. The interrogator is allowed to experience or question only mental attributes. In the next step of the game, a machine is substituted for one of the humans, and the interrogator must differentiate the machine from the human. (*Gregory, 1998, p. 784*) *(19)*

Since there are gender- and machine-based differences in computational capacity, the Turing test differential is within the capacity of many artificial intelligence (AI) systems. If the Turing test is a marker for consciousness, consciousness is within the capacity of intelligent machines.

A mathematician developed the Turing test; the mirror test for assessing the presence of self-awareness was designed by neurobiologists; and the Chinese room test for consciousness was derived in form by the philosopher John Searle, based on the Turing test:

> Imagine that you carry out the steps in a program for answering questions in language you do not understand. I do not understand Chinese, so I imagine that I am locked in a room with a lot of boxes in Chinese symbols (the database), I get small bunches of Chinese symbols passed to me (questions in Chinese), and I look up in a rule book (the program) what I am supposed to do. I perform certain operations on the symbols in accordance with the rules (that is, I carry out the steps of the program) and give back small bunches of symbols (answers to questions) to those outside the room. I am the computer implementing a program for answering questions in Chinese, but all the same I do not understand a word of Chinese. And this is the point: if I do not understand Chinese solely on the basis of implementing a computer program for understanding Chinese, then neither does any other digital computer program solely on that basis, because no digital computer has anything I do not have. (*Searle, 1997, pp. 10–11*) *(20)*

This test argues that in order to have consciousness, a system must demonstrate more than the ability to manipulate formal symbols. A conscious system must also demonstrate mental and semantic content. Daniel Dennett, contending that AI has the capacity for consciousness, has argued that the Chinese room test is confused and misdirected, "full of well conceived fallacies" (21). The philosopher David Chalmers concludes that based on his understanding of the Chinese room model, "The outlook for machine consciousness is good in principle, if not in practice" (22). Colin McGinn, in his review of these tests for consciousness, is more cautionary, indicating that there is little evidence and nothing to explain how the computational complexity of AI or the brain might give rise to consciousness (23).

BEYOND THE CHINESE ROOM

Beyond testing and questioning, and the philosophy of definition, consciousness can be studied using a variety of medical and neuroscientific approaches. Trauma- and disease-induced cognitive deficits led to the first understandings of the anatomy of consciousness. That work was expanded by the use of induced lesions in animal studies. More recently, micropipette systems have been developed that can detect the firing of individual neurons. Techniques for non-invasively assessing neuroanatomy have expanded in recent years beyond X-rays to scanning systems able to provide detailed non-invasive imagery of the functioning CNS. The capacity of drugs to alter consciousness was apparent to ancient shamans. The study of such cognitive drug effects and side-effects evolved into the field of neurochemistry. The electric fields of the brain were first monitored a century ago. Today, complex fields of electrical and magnetic flux can be monitored in the CNS in real time.

PATHOPHYSIOLOGY

Head-injury – induced loss of consciousness led Aristotle to conclude that the seat of consciousness was in the brain (24). Today, the study of brain pathophysiology – whether based on illness, trauma, or surgery – continues to be a useful technique for studying aspects of consciousness, including dreaming (25). This approach has led to the understanding that consciousness, sleep, and dreaming are global states preferentially preserved even after extensive trauma or insult to the brain. Lesion studies, primarily in cats, indicate that REMS is a brainstem-based process that can be eliminated by transecting the brainstem (26). While such lesions in humans are invariably fatal, partial brainstem trauma can affect REMS without having an effect on dreaming. Dreaming, as noted previously, is more likely to be lost after severe bi-basilar frontal CNS damage or after the psychosurgery of frontal lobotomy (27). Unconsciousness can be induced by general blunt forced trauma to the head – a concussion. When more severe, a spectrum of generalized brain insults and trauma can lead to a prolonged vegetative state (coma) in which volitional control and response to external stimuli are varyingly lost for extended periods.

For almost 100 years, X-rays have been used to non-invasively provide increasingly precise pictures of internal anatomy. X-rays, while able to provide insight into skull structure and radio-opaque intrusions (such as bullets), have been of limited use in the study of the brain anatomy. More recently, computed tomography (CT) scans used in concert with injected radio-sensitive markers have provided increasingly detailed descriptions of CNS anatomy. Magnetic resonance imaging (MRI) records the response of magnetically aligned atomic particles to external radio waves. When coupled with the three-dimensional capacity of CT, MRI can provide even more detailed imagery of CNS neuroanatomy. These techniques have allowed physicians to see tiny yet dangerous pathological brain lesions such as vessel aneurysms.

There are newer scanning techniques that can be used to look beyond brain anatomy to brain function. Functional magnetic resonance imaging (fMRI) uses MRI technology to record areas of glucose and oxygen metabolism, lighting up metabolically active areas of the brain. Many newer fMRI studies are blood-oxygen – level dependent, a specificity denoted by the acronym BOLD. The time differential for this approach is short enough that brain activity (function) can be recorded on a real-time basis. Positron emission tomography (PET) records the metabolic uptake of radioactive markers and obtains a lower resolution index of brain activity. These techniques allow the technician to record the brain activity occurring in response to a stimulus or even a thought. Such experiments indicate that complex, widely separated, and individually variable CNS activity occurs in response to experimental stimuli. Experimental stimuli such as visualizing the color red or thinking about God can produce consistent patterns of CNS activation. Specific aspects of consciousness, particularly those in waking, have proven amenable to study, while global aspects such as consciousness have proven more difficult. The brain activity during REMS is quite easy to study using these techniques. Dreaming, a global state, has been particularly difficult to address since it occurs during the uncommunicative perceptual isolation of sleep. Recent studies of lucid dreaming attempt to address this methodological problem and are addressed in Chapter 6.

NEUROCHEMICAL FROG PROPHECIES AND THE CONTROL OF RAPID EYE MOVEMENT SLEEP

If the Lewis-Williams and Dowson shaman's trance hypothesis (see Chapter 2) is correct, the psychedelic plants and fungi that were used to incite the imagery that inspired the cave paintings were among the first neurochemicals utilized by our species. Various milder chewable and smokable neuroactive agents such as betel, coca, caffeine, opiates, and tobacco have a long history of use that has continued into modern society. Archaeologically, the societal development of agriculture marks the advent of ethanol (grain-based alcohol) – also in continued use. Neurocognitive effects and side-effects are among the most common drug actions induced by the many agents in today's pharmacopoeia. Sedation is the most common effect, while stimulation and insomnia are also common. Until recently, these symptoms were viewed as global actions. However, in the past fifty years a wide spectrum of neurotransmitters and neuromodulators have been discovered that affect the activity and intercommunication of nerve cells. The first of these substances was discovered in the 1930s by the Austrian pharmacologist Otto Loewi. He was attracted by the idea that vagal nerve effects on heart rate might be chemically mediated, but had difficulty determining how to prove his postulate. One night, he awakened after a dream in which he was sure that he had discovered the experimental solution. But he was unable to remember his dream. The next night, he went to bed intent on redreaming the solution. When he woke, he rushed to the laboratory, where he electrically stimulated the vagus nerve of a frog to induce a slowing of the heart rate. He took the blood from that frog and injected it into another. The second frog had a slowing of the heart rate as well. This experiment demonstrated that the slowing of the heart rate was mediated by a chemical in the blood. That chemical, acetylcholine, was the first neurotransmitter to be isolated. When, decades later, acetylcholine was discovered to be the primary neurochemical trigger for REMS, it was as if the inherent prophecies present in Loewi's dream had been fulfilled.

Today, acetylcholine has been reclassified. The label "neurotransmitter" is now limited to those chemicals that are required for nerve cells to develop spike potentials. There are only three such neurochemicals: the activators glycine and glutamate and the inhibitor gamma-hydroxy-butyric acid (GABA). The hundred-plus other neuroactive "neurotransmitters" (including acetylcholine) are now classified as neuromodulators, since they modulate the effects of the three basic neurotransmitters (28). Almost all pharmacological agents alter the activity of one or another of these neurotransmitters and neuromodulators affecting CNS alertness, by inducing sedation or stimulation as effects or side-effects. These neurotransmitters and neuromodulators are the chemical modulators of consciousness.

The neurochemistry and neuroanatomy strongly suggest that the state of consciousness is global. The search for a specific neurochemistry of consciousness drugs affecting specific neuroreceptors, like the search for a specific CNS neuroanatomical site of consciousness, is an approach likely doomed to fail.

Dreaming, rather than being affected by those drugs that alter REMS, also turns out to be a neurochemically complex state that, like consciousness, is affected by a wide spectrum of psychoactive agents. Drugs affecting the major neurotransmitters and neuromodulators – dopamine, nicotine, histamine, GABA, serotonin, and norepinephrine (noradrenaline) – alter dreaming and nightmare frequency. Other medications affect dreaming by altering an individual's conscious relationship to the environment (anesthetics) or by affecting the inflammatory response that occurs secondary to infection and illness. Changes in dreaming and nightmares are induced by medications that are known to cause alterations in cognitive arousal. Alerting medications that induce the side-effects of arousal (insomnia), and those that produce the effects or side-effects of sedation are those that affect dreaming and nightmares (29).

THE ELECTROPHYSIOLOGY OF A
STORM-WRACKED OCEAN

After the invention of the electrical amplifier, it was only a matter of time until someone attached electrodes to the human cranium. The first recorded brain electrical activity (in microvolts) was the alpha frequency that occurs with drowsiness and sleep onset. It was first recorded in 1928, when Hans Berger used a string galvanometer attached to scalp electrodes. As near as can be determined, based on remaining records, this discovery had nothing to do with his personal experiences of dreaming. Today, these original two-dimensional single electrode tracings have been expanded to include multiple electrodes that can attain a three-dimensional tracing of the real-time electrical activity of the brain. Electroencephalography (EEG) analysis is used in the clinical setting to assess the presence or absence of seizure disorders, for polysomnographic sleep staging, and to determine the presence or absence of brain neural activity after cerebral insult. The absence of such activity is legally utilized to define the presence of death.

The origin of the EEG remains a topic of debate in the field. Since the discovery that an electrochemical interface at the cell membrane allowed nerve cells to develop spike potentials (fire) and transmit that electric signal over cellular processes and across synapses in transmission-line fashion to the next cell, it has seemed logical that the EEG should reflect the activity of these neurological spike potentials. However, it has been difficult to artificially create EEG electrical activity or to postulate how discrete spike potentials might lead to the propagated synchronous global EEG rhythms that are typical of sleep (30).

Waking EEG, much like a storm-wracked ocean, is characterized by broken waves and asynchronous activity. In the background, however, even the waking EEG includes oscillating electrical waveforms. Oscillatory (synchronous) brain activity is a characteristic CNS property of all mammalian species (31). The Nobel Prize – winning Goldman–Hodgkin–Katz equation (1959) describes our chemical and electrical understanding of nerve cell functioning (32). This electrochemical interface formula describes the physiological basis for a neuron's capacity to develop a spike potential. It also describes how extraneural electrical fields such as the EEG can affect neuron activity. At the neural membrane, neurochemicals and electrical fields are involved in an interconnected dance that affects the tendency of the cell to fire, sets equilibrium states, and supplies energy for cellular functions.

As early as the 1960s, evidence indicated that synchronous extracellular fields such as those described by the EEG could affect the tendency of neurons to fire (33). It is likely that specific physiological frequencies are linked to corresponding metabolic oscillators at the neural membrane based on the rhythmic opening and closing of ionic gateways and channels (potassium for alpha frequency, calcium for sigma frequency) (34). EEG rhythmicity requires many neurons to fire in sequence. Oscillation frequency is modulated by inhibitory neurotransmitter (GABA) activity that affects the activity of the chloride ion channels essential for the synchronized firing of individual neurons in the CNS (35). Improved research techniques, such as brain slice micropipette studies, have provided strong support for the existence of a functioning electrophysiological system that is able to affect the tendency of neurons to fire, reset cellular equilibrium, affect signal-to-noise ratio, act in neural communication, and affect cellular messaging systems (36). EEG synchronization is now viewed by many neuroscientists as a potentially valid mechanism for attaining functional cerebral integration between widely spaced neuronal populations (37). This system, affecting electrically sensitive neuromessaging systems at the neural cell membrane, has the potential to affect expression and access to memory systems, including those memories stored in our electrically sensitive intracellular DNA (38).

Quantitative electroencephalography (QEEG) analysis is a statistical approach that can be applied to multiple-lead EEG systems to describe the occurrence of the different physiological EEG frequencies as well as to localize their CNS location and quantify their relative power. QEEG analysis can map synchronous EEG activity into patterns of EEG topography that demonstrate visually in real time the changing flux of electrical patterns across the CNS.

Magnetoencephalography (MEG) is similar to EEG. It records the magnetic component of the EEG electrical signals of neuronal oscillation. MEG denotes the limits of electrophysiological research, demonstrating that the magnetic potentials of the CNS are highly multivariate, constantly fluxing and changing millisecond to millisecond across the CNS (39). Faster than fMRI and PET, MEG charts the real-time rapid flux of changing synchronous fields in the CNS. Such images are beautiful, compelling, and a bit frightening since they document the temporally changing complexity of the CNS. This is a CNS system of unknown function that just a few years ago was not even known to exist. Even today, there are few neuroscientists cognizant of this system and its potential functional CNS significance.

APPROACHING DREAM TOADS

The study of consciousness is an ideal topic on which to apply Descartes' scientific method. A majority of the attempts to study consciousness have been based on theory, from the top down. As noted in the Preface to this book, neuroconsciousness theories have distracted most scientists from utilizing empirically testable approaches to the study of dreaming and consciousness. In logical sequence we can approach consciousness from Descartes' Step 1 – never accept anything is true that is not known to be so. Since so little is scientifically known about consciousness, it is difficult to even define, so Step 1 is fairly easy. We should dispense with any preconceived notions that we might have about consciousness.

From that baseline, Descartes' Step 2 would be to divide the problem into as many component parts as possible. Luckily, there is a spectrum of forms of consciousness extending from focused wakefulness to the state of living non-consciousness that is dreamless sleep. This spectrum of states is sometimes called the "altered states" of consciousness, with these states altered as compared to focused wakefulness (40). These forms of consciousness, while markedly different from focused wakefulness, are not the psychotic or drug-induced states that are brought to mind by the counterculturally hijacked term "altered states." These are, rather, the states of sleep, rest, and distraction where we spend a majority of our lives. Waking is a variably conscious state comprised of different states of consciousness, with some of these states having dream-like characteristics. The states of sleep qualify as states of consciousness due to the presence of dreaming. Since dreaming occurs with all sleep-associated states, each of the sleep states can be considered a form of consciousness. There is also a multiplicity of states on the borderline of sleep/wake, including focused and unfocused meditation, drowsiness, hypnosis, lucid dreaming, sleep meditation, and the various dream-associated parasomnias occurring on arousal from sleep. These states have dream-like state-specific mentation. From the perspective of consciousness, they have received only limited study. To this point, most researchers and theorists have viewed waking as a generalized state in which attention is focused on the processing of perception. Dreaming has been viewed as a state of consciousness specifically associated with REMS. The borderline states that have been studied, especially lucid dreaming, have been forcibly integrated into this simplified model.

Descartes' Step 3 is to build from the smallest of what is known. Based on recent history, the argument can be made that dream science has received the least scientific study of cognitive states. Since the dream/REMS debacle, dream science is not typically incorporated into the study of consciousness. There is, however, another way to consider "the smallest of what is known." Consciousness can be approached by building down from an analysis of what is known best (focused wakefulness) to the state that includes only the smallest glimmer of consciousness – the remarkable state of dreamless sleep. Each conscious state is composed of multiple components that contribute to that form of consciousness, and each state differs from the other in including some components of consciousness while excluding others. Viewed in this manner, each state can be approached by the analysis of its biological components, as well as by exploring its phenomenology and extent of consciousness. Working from focused wakefulness, consciousness can be dissected into its basic components along the path of a journey toward the barely conscious state of dreamless sleep.

As John Searle points out, "What I mean by 'consciousness' can best be illustrated by examples. When I wake up from a dreamless sleep, I enter a state of consciousness, a state that continues so long as I am awake. When I go to sleep or am put under general anesthetic or die, my conscious states cease. If during sleep I have dreams, I become conscious …" (41). The conscious states, some wake based and some occurring out of sleep, and some on the borderline of waking consciousness, form ideal examples for addressing basic aspects of consciousness. Behaviorally, each state has basic differences in the degree of perceptual isolation, type of thought processing, level of attention, memory access, teachability, and level of conscious control. Since these characteristics have been utilized as criteria for consciousness and since the characteristics of specific conscious states have already been studied, these characteristics can be used to describe the specific form of consciousness associated with each of these states (42). Stanley Krippner, who in the 1960s developed the terminology of "altered states of consciousness," has deconstructed that term and its connotations to coin a new phrase: "patterns of phenomenological properties" (43).

From the perspective of attempting to understand consciousness, these biologically developed examples (forms) of consciousness offer windows that can be used to look into the components of mind and body that comprise consciousness. Much of the underlying neuroscience – the electrophysiology, neuroanatomy, and neurochemistry associated with these states – has been clarified during research in the field of sleep. The empirical, phenomenological, and philosophical understanding of each of these states can be integrated with these physiological data and integrated in a multimodal fashion to explore each of these examples (forms) of consciousness. This dream science–based framework leads to a different perspective of consciousness from the theoretical and sometimes apparently hopeless routes that have been traveled before (44).

In making this journey of exploration into the mind, we should try to comply with the final component of Descartes' method. We should keep a journal of our dreams and carefully keep complete records for review (Step 4). Even reading and thinking about our dreams increase their recall and impact (45). A relative of Loewi's dream toad is likely to be singing in the trees, nursing a lost leg and a slowing heart. There, in that garden, he contemplates the basis of a mortality that lies just beyond our conscious horizon. And then he dreams.

Notes

1. Wilbur, R. (1950) Death of a toad, in *Ceremony and Other Poems*. New York: Harcourt, Brace. Bert States, in his book *Seeing in the Dark* (1997) utilized this, his "favorite" poem, as a metaphor for the dream finch that his sister reported from her final dream before dying. States, a professor of literature at the University of California, Santa Barbara, was one of the first to realize that if we were to make progress in the scientific study of dreams, perspectives outside the scientific specialties were both beneficial and required to facilitate an understanding of the dream state.
2. Sutherland, N. S. (ed.) (1989) *The International Dictionary of Psychology*. New York: Continuum. You have to admire the editor's audacity.
3. Lakoff, G. and Johnson, M. (1980) *Metaphors We Live By*. Chicago, IL: University of Chicago Press. Beyond signifiers, deconstructive logic and philosophy lead eventually to the question of absolute metaphors.
4. Chambers, D. (1996) *The Conscious Mind*. Oxford: Oxford University Press, pp. 3–6.
5. Hobson, J. A. (1996) *Consciousness*. New York: Scientific American Library, p. 16.
6. Duffy, J. D. and Campbell, J. J. (1994) The regional pre-frontal syndromes: A theoretical and clinical overview. *Journal of Neuropsychiatry* 6: 379–387; Fogel, B. S. (1994) The significance of frontal system disorders for medical practice and health policy. *Journal of Neuropsychiatry* 6: 343–347; Pagel, J. F. and Vann, B. (1996) A learning/memory based paradigm describing cognitive integration of dreams: Evidence for such a paradigm derived from reported dream effects on awake behavior in obstructive sleep apneics. *ASD Abstracts*.
7. Lupton, C. (2009) Who in the world: Essay film, transculture and globality, in *Telling Stories: Countering Narrative Art, Theory and Film*, ed. J. Tormey and G. Whitley. Newcastle upon Tyne: Cambridge Scholars., p. 235. Reflexive consciousness is presented in most neuroconsciousness texts as a self-defined concept. It is interesting that this more reasoned and insightful paradigm for the concept is to be found in the area of film and narrative study.
8. McGinn, C. (1991) *The Problem of Consciousness*. Oxford: Blackwell. Chambers, D. (1996) *The Conscious Mind: In Search of a Fundamental Theory*. New York: Oxford University Press.
9. Roitblat, H. (1987) *Introduction to Comparative Cognition*. New York: Freeman.
10. Shettleworth, S. (1972) Constraints on learning, in *Advances in the Study of Behavior*, ed. D. Lehrman, R. Hinde and E. Shaw. New York: Academic Press.
11. Griffin, D. (2001) *Animal Minds: Beyond Cognition to Consciousness*. Chicago, IL: University of Chicago Press, pp. 23–25.
12. Patterson, F. and Cohn, R. (1994) Self-recognition and self-awareness in lowland gorillas, in *Self-Awareness in Animals and Humans: Developmental Perspectives*, ed. S. Parker, R. Mitchell and M. Boccia. New York: Cambridge University Press.
13. Tomascello, M. and Call, J. (1997) *Primate Cognition*. New York: Oxford University Press.
14. Gineste, M. (1997) *Analogie et Cognition*. Paris: Presses Universitaires de France, p. 119.
15. De Beaune, S. (2009) in *Technical Invention in the Paleolithic: What if the Explanation Comes from the Cognitive and Neurophysiological Sciences?*, ed. S. de Beaune, F. Coolidge and T. Wynn. Cambridge: Cambridge University Press, pp. 3–14.
16. Dement, W. and Vaughan, C. (2000) *The Promise of Sleep*. New York: Dell.
17. Damasio, A. (1999) *The Feeling of What Happens: Body and Emotion in the Making of Consciousness*. New York: Harcourt Brace, p. 100.
18. Valli, K., Frauscher, B., Gschliesser, V., Wolf, E., Falkenstetter, T. and Schönwald, S. V., et al. (2011) Can observers link dream content to behaviors in rapid eye movement sleep behavior disorder? A cross-sectional experimental pilot study. *Journal of Sleep Research* 21: 21–29. Katja Valli, the purple-haired Finn, is doing some of the best of current work in the field of dream science.

19. Turing, A. (1998) in *The Oxford Companion to the Mind*, ed. R. Gregory. Oxford: Oxford University Press, p. 784.

20. Searle, J. (1997) *The Mystery of Consciousness*. New York: A New York Review Book, pp. 10–11.

21. Dennett, D. (1991) *Consciousness Explained*. Boston, MA: Little, Brown and Co., pp. 438–440. Dennett, D. (1997) An exchange with Daniel Dennett, in *The Mystery of Consciousness*, ed. J. Searle. New York: A New York Review Book, p. 116.

22. Chalmers, D. (1996) Strong artificial intelligence, in *The Conscious Mind: In Search of a Fundamental Theory*. New York: Oxford University Press, pp. 322–331.

23. McGinn, C. (1991) *The Problem of Consciousness*, pp. 212–213.

24. Van Der Eijk, P. (2005) *Medicine and Philosophy in Classical Antiquity: Doctors and Philosophers on Nature, Soul, Health and Disease*. New York: Cambridge University Press.

25. Solms, M. (1997) *The Neuropsychology of Dreams – A Clinico-Anatomical Study*. Mahwah, NJ: Lawrence Erlbaum.

26. Jouvet, M. (1962) Recherches sue les structures nerveuses et let mecanismes responsables des differentes phases du sommeil physiologique. *Archives Italiennes du Biologie* 100: 125–206. It is Jouvet who, with Dement, began the modern scientific study of sleep. I once heard him give a rather remarkable lecture describing how neural cells in the nose are activated during REMS at the same time as the penile/clitoral erection typical of the state.

27. Solms, M. and Turnbull, O. (2002) *The Brain and the Inner World: An Introduction to the Neuroscience of Subjective Experience*. New York: Other Press, pp. 141–143.

28. Kandel, E. and Siegelbaum, S. (2000) in *Synaptic Integration in Principles of Neuroscience*, 4th ed. E. Kandel, J. Swartz and T. Jessell. New York: McGraw Hill, pp. 207–228.

29. Pagel, J. F. and Helfter, P. (2003) Drug induced nightmares – An etiology based review. *Human Psychopharmacology: Clinical and Experimental* 18: 59–67; Pagel, J. F. (2006) The neuropharmacology of nightmares, in *Sleep and Sleep Disorders: Neuropsychopharmacologic Approach*, ed. S. R. Pandi-Perumal, D. P. Cardinali and M. Lander. Georgetown, TX: Landes Bioscience, pp. 241–250. Pagel, J. F. (2008) Sleep and dreaming – Medication effects and side-effects, in *Sleep Disorders – Diagnosis and Therapeutics*, ed. S. R. Pandi-Perumal, J. C. Verster, J. M. Monti, M. Lader and S. Z. Langer. London: Informa Healthcare, pp. 627–642.

30. Christakos, C. N. (1986) The mathematical basis of population rhythms in nervous and neuromuscular systems. *International Journal of Neuroscience* 29: 103–107; Buzsaki, G. (2006) *Rhythms of the Brain*. Oxford: Oxford University Press.

31. Cantero, J. and Atienza, M. (2005) The role of neural synchronization in the emergence of cognition across the wake–sleep cycle. *Reviews in the Neurosciences* 16: 69–84.

32. Hodgkin, A. L. and Horowicz, P. (1959) The influence of potassium and chloride ions on the membrane potential of single muscle fibers. *Journal of Physiology* 148: 127–160.

33. John, E. R. and Swartz, E. L. (1978) The neurophysiology of information processing and cognition. *Annual Review of Psychology* 29: 1–29. This early paper was far ahead of its time. It is only now that researchers are waking up to the importance of the neural chemical/electrical interface.

34. Cheek, T. R. (1989) Spatial aspects of calcium signaling. *Journal of Cellular Science* 93: 211–216. Steriade, M. (2001) *The Intact and Sliced Brain*. Cambridge, MA: MIT Press.

35. Liu, C. and Reppert, S. M. (2000) GABA synchronizes clock cells within the suprachiasmatic circadian clock. *Neuron* 25: 123–128; Buzsaki, G. (2006) *Rhythms of the Brain*.

36. Anastassiou, C., Perin, R., Markam, H. and Koch, C. (2011) Emphatic coupling of cortical neurons. *Nature Neuroscience* 14: 217–223; Axmacher, N., Mormann, F., Fernandez, G., Elger, C. and Fell, J. (2006) Memory formation by neuronal synchronization. *Brain Research Reviews* 52: 170–182; Cash, S., Halgren, E., Dehghani, N., Rossetti, A., Thesen, T. and Wang, C., et al. The human K-complex represents an isolated cortical down

state. *Science* 324: 1084–1087; Steriade, M. (2006) Grouping of brain rhythms in cortico-thalamic systems. *Neuroscience* 137: 1086–1106.

37. Singer, W. (1999) Neuronal synchrony: A versatile code for the definition of relations? *Neuron* 24: 49–65; Salinas, E. and Sejnowski, T. (2001) Correlated neuronal activity and the flow of neural information. *Nature Reviews. Neuroscience* 2: 539–550; Cantero, J., Atienza, M., Stickgold, R., Kahana, M., Madsen, J. and Kocsis, B. (2003) Sleep dependent theta oscillations in the human hippocampus and neocortex. *Journal of Neuroscience* 23: 10897–10903.

38. Pagel, J. F. (1990) Proposing an electrophysiology for state dependent sleep and dream mentation. *APSS Abstracts*, 134; Pagel, J. F. (2012) The synchronous electrophysiology of conscious states. *Dreaming* 22: 173–191. This perspective, initially derided as pseudoscience, is now cutting edge for the fields of cognitive science.

39. Besserve, M., Jerbi, K., Laurent, F., Baillet, S., Martinerie, J. and Garnero, L. (2007) Classification methods for ongoing EEG and MEG signals. *Biological Research* 20: 415–437; Kiebel, S., Garrido, M., Moran, R., Chen, C. and Friston, K. (2009) Dynamic causal modeling for EEG and EMG. *Human Brain Mapping* 30: 1866–1876.

40. Krippner, S. (1972) Altered states of consciousness, in *The Highest States of Consciousness*, ed. J. White. Garden City, NY: Anchor Books., pp. 1–5.

41. Searle, J. R. (1998) *The Rediscovery of the Mind*. Cambridge, MA: MIT Press, p. 83.

42. Pagel, J. F. (2012) The synchronous electrophysiology of conscious states.

43. Rock, A. and Krippner, S. (2007) Shamanism and the confusion of consciousness with phenomenological content. *North American Journal of Psychology* 9: 485–500.

44. Crick, F. and Mitchison, G. (1983) The function of dream sleep. *Nature* 304: 111–114.

45. Strauch, I. and Meier, B. (1996) *In Search of Dreams: Results of Experimental Dream Research*. Albany, NY: University of New York Press.

5

The Forms of Waking Consciousness

*Then I turned my attention to the study of some arithmetical questions appar-
ently without much success and without a suspicion of any connection with my
previous researches. Disgusted with my failure, I went to spend a few days at
the seaside, and thought of something else. One morning, walking on the bluff,
the idea came to me, with just the same characteristics of brevity, suddenness,
and immediate certainty, that the arithmetic transformations of intermediate
ternary quadratic forms were identical with those of non-Euclidean geometry.*
(Poincaré, 1929) (1)

When awake, we are constantly thinking. We think in much of our
sleep. Thinking can be viewed as the diverse, interrelated central nervous
system (CNS) that functions to process thought content. Thought content
is different from thought processing (the acts of receiving, perceiving, comprehending, storing, manipulating, monitoring, controlling,

Dream Science.
DOI: http://dx.doi.org/10.1016/B978-0-12-404648-1.00005-3

TABLE 5.1   Types of Rational Thought

Analytical	Comparing and contrasting, evaluating, reason, logic, reflection, contemplation	27.2%
Perceptual	Paying attention to visual, auditory, gustatory, olfactory, tactile, and kinesthetic occurrences	22.4%
Memory and time awareness	Remembering and recall; recognition of characters, history, events, abilities, time, and the dreaming state	15.0%
Affective	Distinguishing, naming, and/or verbalizing the spectrum of feeling and psychophysiological states	9.5%
Executive	Decision making, problem solving, planning and agency	7.3%
Subjective	Personal history, characteristics, appearance, beliefs and desires, skills and goals	7.3%
Intuitive/ projective	Assumption, lack of sufficient facts, erroneous attributions	6.4%
Operational	Reading, writing, counting, measuring	4.7%

Note: The numbers in the last column denote percentages of these thought types present in dream reports (4).

and responding to data) – what most of us refer to as "thinking" (2). Behaviorally, waking is characterized by the integration, utilization, and response to sensory input. The analysis of perceptually based thought content (all data received and accumulated by the individual) occurs primarily during waking. During focused waking, thought processing is at its most exact, a precision based on the requirements, constraints, and interactive rules of waking life. Waking thought is more "rational" than dreaming thought, with rational thought being the individual's ability to notice and differentiate aspects of personally subjective experience from known objective data (3). The series of categories of thought process associated with rational thought are listed in Table 5.1. Waking thought, like dream thought, is varyingly rational.

Although heavily studied and tested as an aspect of functional performance, waking is the most poorly classified area of consciousness. Fifty years ago we understood almost nothing of sleep, but scientific emphasis and study have since been aggressively applied. The analysis of waking has received far less emphasis, and now we find ourselves in the logically strange situation in which waking is generally viewed as a poorly understood "black box" of cognitive processing during which perceptual input takes place. Almost all psychological and educational studies and tests have concentrated on the analysis of capabilities present during the state of focused waking. Yet, waking varies throughout the day, with different processes of thought, levels of focus and alertness, memory access, and sense of time associated with different types of waking (5). During waking, there is a wide variety of "non-attentive" conscious states in which the available perceptual input is de-emphasized. We spend a majority of our waking time in these non-focused states.

ATTENTION-FOCUSED WAKING

Our educational system can be viewed as a training program for the establishment and maintenance of the conscious state of focused waking. Intelligence quotient (IQ) tests have the remarkable capacity to predict school performance and later life success in Western society (6). IQ tests assess the capacity to recall stored knowledge – what is sometimes called "crystallized intelligence." These tests of intelligence also test for the students' ability to attain and maintain a state of focused waking. This ability to reason or solve problems when applied to novel domains is called "fluid intelligence" (7). Focused waking often includes the various types of rational thought processing (Table 5.1).

Attention, as an aspect of focused waking consciousness, has proven amenable to experimental study. Our conceptual and implicit understanding of attention has changed somewhat since the topic was addressed in 1890 by William James:

> Everyone knows what attention is. It is the taking possession by the mind, in a clear and vivid form, of one out of what seem to be several simultaneously possible objects or trains of thought. Focalization, concentration of consciousness are of its essence. It implies withdrawal from some things in order to deal effectively with others. (James, 1890) (8)

Attention has at least three components: executive attention, alerting, and orienting. What James described as attention is now called "executive attention," the component of attention involved in focusing and narrowing the focus of attention. Executive attention is anatomically based in the anterior cingulate gyrus of the frontal cortex. Alerting is a rapid response to perceptual and sense input, characterized in psychological testing by an increased error rate compared to executive attention. Alerting is neuroanatomically associated with the right medial aspect of the frontal lobe. Orienting, the process of positioning and placement with reference to time and place, takes place in the anterior occipital/posterior parietal region, the most posterior of these CNS locations (9).

Beyond the behavioral aspects of attention, quantitative electroencephalography (QEEG) has been useful in classifying and comparing the waking conscious states. Several authors have proposed that consciousness itself has an electrophysiological signature in the beta/gamma frequency that occurs during cognitive states of attention (10).

We live in the pervasive electrical grid of 60 Hz alternating electrical currents. In our modern hospitals, background electrical activity can only be eliminated with the use of filters that technically eliminate the beta and gamma (30–80 Hz) EEG frequencies that are in the same physiological range. The experimental definition of consciousness as attention, coupled with the proposed electrophysiological correlate for consciousness, has led some authors to suggest that the electrophysiological correlates of attention be utilized to define consciousness and model consciousness states (11). Short-lived but highly coherent oscillation in the gamma/beta frequency has been reported to occur during the processing of perceptual input during focused waking (12). Gamma/beta is postulated to act as a "binding" frequency with the functional ability to produce a temporal synchronicity of neurons involved in a specific cognitive process (13). Beta/gamma oscillations are potentially involved in the process of "working memory," the type of memory that enables us to keep non-current memories available for cognitive processing (14).

The conflation of attention with consciousness has been in some ways unfortunate. Consciousness is much more than attention, and many attentional processes (e.g., where we move our eyes when we focus our attention) are unconsciousness (15). Attention is sensorally multi-faceted, with each of these perceptual systems (visual, auditory, tactile, and olfactory) controlled by an associated system for perceptual processing. Parallel but separate systems executively control those processes of attention (16). The multiple processing systems for attention and their behavioral expression are under central cognitive control. Through central attention each of these systems is allocated to handle competing demands of information processing (17). Since this is a book focusing primarily on dream science, focused waking – the least dream-like of cognitive states – is considered primarily as a comparative marker for other conscious states (Table 5.2).

DEFAULT NON-ATTENTIVE WAKING

Only a portion of our waking is spent in focused awareness of our external environment. When left undisturbed, the waking thought process is often internally oriented to what is sometimes described as "mind wandering": tasks such as autobiographical memory retrieval, envisioning possible futures, and conceiving the perspectives of others (18). The brain regions involved in those processes that become activated when goal-directed cognitive activity ceases are included as part of a neuroanatomical region described as the default network. The default network includes medial prefrontal regions of the frontal cortex, areas activated during self-referential cognitive processing (19). In addition to this area,

TABLE 5.2 Comparative Behavioral and Synchronous Electroencephalographic (EEG) Characteristics of Waking States

	Perceptual Isolation	Thought-Attention	Associated Synchronous EEG Frequencies
Focused waking	Lowest	Focused	Beta/gamma
Self-referential (default) waking	Low	Self-focused	Beta/gamma, delta, 0.5 Hz activity
Drowsy waking	Low	Unfocused	Alpha
Creative waking	Low–moderate	Focused	NS
Hypnosis	Moderate	Variable	NS
Focused meditation	Moderate	Focused	Gamma, alpha
Unfocused meditation	Moderate	Focused/unfocused	Alpha, theta

Note: NS = not studied.

mind wandering recruits areas of the frontal executive network during periods of decreasing external task demands. These anatomical areas are active during large portions of waking. When asked to complete a prescribed and focused task, most individuals will report mind wandering and being off task more than forty percent of the time. Mind wandering is a global process that is most active during the periods of relative rest that take place after times of intense focused waking (20).

Default network activities such as self-awareness and daydreaming are reported to be less positive and vivid among psychiatric inpatients (21). In children, changes in default network activity are associated with the diagnoses of attention deficit/hyperactivity disorder (ADHD) and sleep apnea (22). Decreased default network activity has been documented in patients with diagnoses of schizophrenia, autism, and Alzheimer's disease (23). In individuals with waking anxiety, default network activity may be compromised. Anxiety states are at least in part secondary to a subject's inability to self-reflect and attain relative CNS rest during waking. Anxiety is a cognitively disorganized state that has significant negative effects on executive task-based capabilities such as creative performance and divergent thinking, processes that require the integration of default system activity (24).

The treatment indications for psychoactive medications are inferred based on their neurochemical class-specific effects on a variety of psychiatric diseases. The effects of psychoactive drugs are likely to be secondary to their effects on specific neurotransmitters (see Chapter 4). Many neurotransmitters and neuromodulators affect neurochemical activity in the default system. While psychoactive neurochemicals exert their effects in controlling the symptoms and expression of illness, the actual manner in which those agents affect such major disease processes as epilepsy, schizophrenia, and mood disorder/depression is still not clearly understood. Psychoactive medications can be classified according to potential function based on their tendency to induce declines and/or increases in specific EEG frequencies (Table 5.3). It is likely that the actions that these agents exert on the default network electrophysiological systems lead to many of the psychoactive effects that these agents exert in treating psychiatric and neurological diseases (26).

The default anatomical area has functional coherence based on EEG topography mapping of the alpha frequency – the predominant CNS electrophysiological frequency (27). Low-frequency fluctuations at very slow frequencies (0.5–1.5 Hz) are also associated with default network activity. Infrared spectroscopy and positron emission tomography (PET) scanning indicate that changes in metabolic and hemodynamic activity occur in association with this electrophysiological activity (28). These synchronous electrical frequency systems are integrated, and during waking coalesce into a system that is active during unfocused and internal modes of cognition (29). These same EEG rhythms are active during most of sleep (30). During these episodes, many neurons from diverse and anatomically disconnected parts of the brain are firing in sequence, constantly fluxing, and changing millisecond to millisecond across the brain (31). Metabolic brain activity fluctuates in association with these rhythms, so there is little question that actual neural brain activity is taking place where this electrical rhythmicity is present (32). When these synchronous EEG rhythms occur, the CNS default system actively engages in processes of internal mentation – the introspective and adaptive mental activities in which humans spontaneously and deliberately engage every day (33).

The default network is likely to be important for our ability to maintain high-quality waking function. Less obviously than focused waking, we may need the activity of the default network for optimal cognitive processing. Our schooling system rarely trains students in mind wandering. It is rather a capacity likely to result in a diagnosis of learning or attention deficit and treatment with activating medications such as methylphenidate (Ritalin) or dextroamphetamine (Dexedrine). Our society currently suffers from an epidemic of these diagnoses in children as well as a rapidly increasing frequency in the diagnoses of anxiety disorders in adults. Focused waking is improved when coupled with episodes of default network activity (brain activity in a different mode). Educational

TABLE 5.3 Consistent Quantitative Alteration in Physiological Electroencephalographic Frequencies Induced by Different Classes of Psychoactive Medication (25)

| | Delta | Theta | Alpha | Sigma | Beta/Gamma | |
	0.5–1.5 Hz	5.5–8.5 Hz	8.5–11 Hz	12–16 Hz	21–32 Hz	Treatment Indications
Benzodiazepines	Alteration		Decrease	Increase	Increase	Anxiety, sedation
Barbiturates				Increase	Increase	Anxiety, sedation, epilepsy
Tricyclic antidepressants	Decrease	Decrease			Increase	Depression
SSRI antidepressants	Decrease		Increase			Depression, anxiety, obsessive-compulsive disorder
Amphetamines	Decrease	Decrease			Increase	Somnolence, narcolepsy, ADHD
Opiates	Increase		Decrease			Pain
Anticonvulsants: phenytoin, valproate, carbamazepine		Increase				Epilepsy
Gamma hydroxybutyrate	Increase					Cataplexy, narcolepsy
Classic neuroleptics		Increase	Decrease	Decrease		Schizophrenia, psychosis

Note: SSRI = selective serotonin reuptake inhibitor; ADHD = attention deficit hyperactivity disorder.

approaches proposing alternatives to the current system of emphasis on the maintenance of focused waking are likely to be of benefit for some individuals.

THE RHYTHMS OF CREATIVITY

Creative waking is not focused waking. Creative insights occur most often during the periods of non-attention that occur after periods of focused waking (see the Poincaré quotation at the start of this chapter). During tests of waking problem solving, individuals given a four-hour task break were more likely to solve the problem (85 percent) than those given no break (55 percent) and those given only a thirty-minute break (64 percent) (34). During the periods of non-task focus, individuals were able to drop non-productive approaches to the problem, and on return take a fresh and more successful approach. This process of waking incubation works best when the periods of non-focused waking follow periods of intense focus on a particular problem (35). Functional magnetic resonance imaging (fMRI) and magnetoencephalography (MEG) data suggest that during such creative processing both the executive attention and the non-focused default regions of the CNS are activated. During optimal task performance there is improved functional connectivity between these two very different areas of CNS function (36).

Daydreaming and waking fantasies occur during non-attentive waking. For most individuals, daydreams (defined as thoughts unrelated to current tasks) involve problem solving and the planning of future activities, except in young males, whose daydreams are most likely to focus on aspects of love and sex (37). Daydreaming includes non-perceptual visual imagery (38). Artists and non-artists alike use the intense imaginative activity of daydreaming in creative activities (39). Some writers and visual artists report that their creative process is much like being in a waking dream. Many report that creative waking has the dream-like characteristics of a lack of awareness of the external world as well as a loss of the sense of time (40).

The major EEG brain rhythms occur at frequencies that can be physically accomplished. Delta frequency occurs at one cycle per second. This is an easily attainable bass drum beat. The maximal attainable rate for finger tapping for most individuals is ten taps per second, equivalent to the EEG dominant alpha 10 Hz frequency (41). Such rhythmic behavior affects the EEG frequency-based rhythms. When the tasks are rhythmic in nature, improved task performance and focus are more likely to occur (42). The improved task performance occurs in association with the development of the EEG rhythms in the CNS (e.g., professional dancers demonstrate a higher degree of alpha synchronization during imagined dance than do novice dancers) (43). Increased EEG alpha activity and synchronization of that activity across the CNS occur during creative processing (44). EEG neurofeedback at alpha and theta frequencies has been used to help individuals to attain optimal dance and music performance (45).

There is a long history of the use of the creativity-based therapies of art, dance, and drama in the treatment of psychiatric illnesses (46). As noted, these creative processes, particularly when rhythmic, can affect the default network. These approaches are likely to allow for an integration with the creative aspects of the various forms of sleep/dream consciousness (Chapter 6). Frequency-based therapies including eye movement desensitization and reprocessing (EMDR) and EEG frequency neurofeedback are known to positively affect psychological functioning (47). Enhancing the electrophysiology associated with CNS default network activity can potentially enhance an individual's capacity to maintain relaxed non-attentive consciousness during waking. Conceptually, rhythmic creative activity and episodic mind wandering between periods of focus are approaches that can be used to provide an excellent framework to assist in fostering creativity. There are also cogent suggestions that such techniques are useful in the psychiatric treatment of anxiety disorders.

THE COGNITIVE CURIOSITY OF HYPNOSIS

Hypnosis is a somewhat strange cognitive process that most of us have not objectively experienced. The participant is guided by a hypnotist to respond to suggestions for changes in sensations, perceptions, thoughts, feelings, and behaviors (48). Its history may be as old as that of shamanism (49). It has been suggested that the trance stages of shamanic consciousness described by Dowson and Lewis-Williams (see Chapter 2) may actually be stages of shamanistic-induced hypnosis (50).

The primary characteristic of the hypnotic state is the loss of self-conscious source monitoring, sometimes called "reality testing" (51). Hypnosis, like sleep, is characterized by a suspension of peripheral awareness. Unlike sleep, hypnosis is a state of highly focused attention (52). Subjects of hypnosis are unable to discern that their induced sensations are imaginal in source. In this fashion, hypnosis resembles dreaming sleep. Insomniacs can be taught self-hypnosis in order to relax before initiating sleep. Hypnosis has also been used to assist in the induction of lucid dreaming (53).

Neuroimaging studies of the hypnotic state using PET and fMRI indicate that attentional components of the frontal cortex are activated during hypnosis, the same areas involved in the processing of cognitively demanding or emotionally salient tasks. These are the anterior regions (the focusing components) of the attentional system. Phenomenology suggests that the state of hypnosis should involve interactions between the prefrontal cortex, the anterior cingulate gyrus, and the sensory processing systems of the CNS; however, the newer brain imaging techniques have been applied only minimally to the study of hypnosis (54).

The EEG changes seen with hypnosis resemble those seen at sleep onset and with meditation states (55). Despite attempts using a wide spectrum of sampling, measuring, and instrumental methodologies, no consistent pattern of electrophysiological changes has been identified for either the hypnotic state or hypnotic tendency. More recent approaches have failed to support the proposal that hypnosis is associated with 75–120 millisecond microstates (56). Hypnosis, today, is often viewed as a cognitive curiosity that was part of the behavioral era of psychology. It is a waking state that co-opts our systems of attentional focus in an unusual way. It is unclear how hypnosis, a waking state that we understand only minimally, affects our normal waking functioning.

MEDITATIVE STATES

Meditation has proven to be a difficult topic to study. Since there are many different approaches and techniques, there is no typical phenomenology associated with meditative practice. Meditation is part of many different cultures and part of almost every religious tradition (57). In the Hindu and Buddhist traditions, meditation is incorporated into the basic structure of the religion. In some traditions, meditation is practiced as an intense cognitive study that includes empirical analysis of the associated cognition:

> The apparently simple act of seeing a rose, for example, is in reality a very complex process composed of different phases, each consisting of numerous smaller combinations of conscious processes, which again are made up of several single movements of consciousness following each other in a definite sequence of diverse functions. Among these phases there is one that connects the present perception of a rose with a previous one and there is one that attaches the present perception of the name "rose" remembered from previous experiences. *(Thera, 1998) (58)*

Some meditative traditions have extended this analysis, approaching consciousness in philosophical detail:

> Consciousness is actually a continuum of momentary acts of awareness that rise and fall along with their respective objects due to specific causes and conditions …

Consciousness acts as a type of motion detector, which when triggered, simply illuminates any objects that are within its range ... Each moment of consciousness can be classified according to the mental factors with which it is associated. *(Flickstein, 2007) (59)*

While some meditative approaches focus on such an intense intellectually based cognitive integration, many others take an experiential approach. Meditation can include focusing on an image, the breath, or a repetitive sound. Other traditions, such as Zen Buddhism, attempt to avoid cognitive processing during meditation. Mindfulness meditation, generally a Vipassana Buddhist technique, is based on focusing attention on the breath during meditation. Brain scanning studies indicate that experienced meditators have activation in brain regions also involved with attention and introspection – some of the same areas activated in hypnosis (60). fMRI studies indicate that mindfulness meditation affects aspects of the medial prefrontal cortex that are involved in the functioning of the default network (61).

Few attempts have been made to achieve comparative analysis of different meditation types and approaches. While there have been efforts to apply Western analytical and scientific tools to the study of meditation, most of these approaches have been used to analyze meditation within a single traditional approach. This is due, in part, to the extensive training required for developing proficiency in even one of these techniques. Within each tradition there is a marked difference in individual proficiency and meditative capacity. Training and duration of the meditative practice affect how that practice affects meditative capacity and experience as well as CNS functioning.

It has been difficult to determine whether there are typical electrophysiological effects associated with meditative practice. Although EEG frequency changes are associated with long-term meditative practice, these changes vary based on meditative technique (62). Long-term Buddhist practitioners have a higher ratio of beta/gamma (25–42 Hz) band activity to theta/alpha activity (4–13 Hz) (63). Some types of Hindu and Tibetan Buddhist yoga have been shown to induce high-frequency biphasic ripples of gamma waves during meditation (64). The pattern of beta/gamma (33–44 Hz) band activity differs in focused meditators from that found in individuals practicing unfocused meditation styles (65). Increases in frontal alpha/theta activity are demonstrated by experienced Zen meditators (66). Studies of meditation-induced relaxation have demonstrated an increase in frontal midline theta (4–8 Hz) activity compared with subjects relaxing while listening to music (67). Increased frontal theta and posterior alpha are apparently characteristic of unfocused meditative techniques (68). This confusion of data indicates that meditation alters CNS electrophysiology, and that this change in electrophysiology occurs in a variety of different patterns based on meditative technique.

Meditation, like hypnosis, is sometimes used as a behavioral treatment to teach relaxation before initiating sleep. As noted previously (see Chapter 1), individuals from both Hindu and Buddhist traditions with extensive training in meditation may sleep little and lose the ability to dream (69). In some Tibetan Buddhist traditions, meditative practice is extended into sleep so that it becomes unclear whether the meditating individual is meditating in sleep or while awake (70).

THE ALTERED STATE OF DROWSY WAKING

Daytime sleepiness is a very common state of consciousness, but it is most often addressed as a pathological state. Daytime sleepiness, generally considered to be "the subjective state of sleep need," interchangeably includes drowsiness, languor, inertness, fatigue and sluggishness (71). More than twenty percent of American adults report excessive daytime sleepiness (EDS) interfering with their daily activities (72). EDS is most common in adolescents, elderly people, and shift workers (73). The actual assessment of true incidence is confused by variations in the commonly applied terminology. In Western medical practice, fatigue (etymologically the French word for sleepiness) is clinically differentiated from sleepiness in diagnostic assessment.

The actual incidence of EDS is also confounded by a lack of correlation between the various techniques used to assess whether it is present. Sleepiness has variable effects on waking function. Sleepy individuals perform the worst when attempting boring and repetitive tasks. With sleepiness a factor in over twenty percent of motorway accidents, EDS is a significant problem on the highways (74). According to the National Transportation Safety Board, approximately fifty percent of one-vehicle accidents involving heavy trucks are fatigue related, with the driver reporting falling asleep in more than seventeen percent of crashes (75). Adolescents and young adults are some of the sleepiest people in our society. A majority of falling-asleep crashes occur in this age group. Sleepy adolescents demonstrate significantly lower school performance, increased tardiness, and lower graduation rates than their non-sleepy classmates (76). The three most common causes for EDS are sleep deprivation, the sleep-associated diagnosis of obstructive sleep apnea (OSA), and the use of commonly prescribed medications and drugs of abuse that induce daytime sleepiness (77). Sleepiness is usually assessed by questionnaire (Box 5.1). Sleepiness can also be assessed using tests of reaction and coordination or tests that assess complex behavioral tasks likely to be affected by sleepiness (e.g., tests of driving performance) (78). Since performance measures are susceptible to the influences of motivation, distraction, and comprehension of instructions, and since sleepiness varyingly affects different waking behaviors, the results obtained from performance tests may differ markedly from those obtained using questionnaires (79).

The drowsy waking state occurring just before sleep onset is behaviorally associated with dream-like episodes of mind wandering and non-perceptual imagery. Waking hallucinations can affect up to eighty percent of subjects during the periods of drowsiness that occur after experimentally induced sleep deprivation (80). In ninety percent of humans, drowsy waking is electrophysiologically marked by the presence of the eyes-closed alpha rhythm. The onset of sleep is defined by a drop in alpha frequency to less than fifty percent of the EEG recording. In ten percent of individuals (those without coherent alpha), sleep onset occurs with a combination of slowing EEG activity, slow rolling eye movements, vertex sharp waves, rhythmic theta activity, hypnogogical hypersynchronicity and/or high-amplitude 3–5 Hz activity (81). These individuals function normally without exhibiting coherent alpha – the predominant electrophysiological rhythm of the CNS. Drowsiness is one of the most commonly experienced altered forms of waking consciousness. Many of us, owing to medical/psychiatric disease and/or medication, may spend a majority of our waking lives in this altered state.

FORMING WAKING CONSCIOUSNESS

Waking is often viewed as a global state in which perceptions are integrated and acted upon. It is clear, however, that waking consists of multiple and often discrete states of consciousness. These waking states share the overall characteristic of perceptual access, yet they can have very different phenomenologies. Some are states of focus; others are states of relaxation from that focus; some are states of control, and others, such as hypnosis, are states in which control systems seem to be bypassed.

Primary forms of consciousness involve the awareness and integration of perception and sensations, and the resultant motor responses. These systems are amenable to neuroanatomical study. In both humans and animals, specific neuroanatomical damage to these systems leads to consistent cognitive deficits. Scanning techniques can be used to delineate the specific brain areas that are involved in primary consciousness processing of perceptions, sensory data, and motor activity.

In humans, secondary and tertiary conscious processing also takes place during all forms of waking consciousness. All of the described forms of waking consciousness have non-perceptually based components. During waking, episodic segments of non-perceptual, non-focused consciousness allow for alternative processes of internal cognition. These alternative processes can occur during attentional focus, but are most often present during periods of waking recovery from focus, associative thought, imagery, and alternative approaches to problem solving and periods of drowsiness that can lead to the states of sleep. The tertiary/ executive processes of consciousness, including self-reflective awareness, planning, organization, execution of complex goal-directed behavior, flexible response to changing environmental contingencies, dream incorporation into waking behavior, task persistence despite distraction, and creative problem solving, occur periodically during waking. These global, complex processes involve large and widely separated components of the CNS. Neuroanatomically and neurochemically, it has been difficult to describe specific brain sites or brain chemicals that affect tertiary conscious processes.

Electrophysiology, initially applied to sleep, has proven more useful. When internal, non-perceptually based cognition is taking place, the basic form of the EEG changes. The disordered and agitated EEG of perception is integrated or replaced by synchronous waveforms occurring at the different physiological frequencies. Attention in focused waking is marked by beta/gamma activity. During default waking very slow subdelta activity develops, as well as the runs of alpha and theta activity that characterize hypnosis, meditation, and drowsiness. During waking, these rhythms rapidly flux and change across the CNS. This is not transmission-line processing based on neurons communicating through their neural processes or an anatomically correlated and defined process. Something quite different is taking place in the CNS. A global system is operating that affects the tendency of neurons to fire, resets chemical and electrical equilibrium, affects signal-to-noise ratio, acts in neural communication, and affects internal cellular messaging systems (82). This system of electrical synchronization is utilized to attain functional cerebral integration between widely spaced neuronal populations, and to affect electrically sensitive neuromessaging systems at the neural cell membrane (83). These extracellular electrical fields can affect expression and access to memory systems, including those stored in electrically sensitive proteins such as cellular DNA (84). There is good evidence that these systems are involved in and required for the processes of tertiary consciousness (85).

This system expresses itself even more fully during sleep when perceptual and motor activity is blocked. When this electrophysiological system is active during waking, waking consciousness assumes dreamlike aspects. Perception, time sense, and orientation become secondary and less active components of the waking experience. During these episodes, associative thoughts and memories, visual imagery, and even remembered dreams become part of waking. At the extreme, these states

BOX 5.1

EPWORTH SLEEPINESS SCALE

Rate the chance that you will doze off in the following situations:

0 = No chance of dozing
1 = Slight chance of dozing
2 = Moderate chance of dozing
3 = High chance of dozing

Sitting and reading ____
Watching TV ____
Sitting inactive in a public place such as a theater or meeting ____
As a passenger in a car riding for an hour without breaks ____
Lying down in the afternoon when circumstances permit ____
Sitting and talking to someone ____
Sitting quietly after lunch without alcohol ____
In a car while stopped for a few minutes in traffic ____

+   _ _ _ _ _ _ _

Add above for total score ____
Less than 8 – Indicates normal daytime alertness
8–11 – Indicates mild sleepiness
12–15 – Indicates moderate sleepiness
16–24 – Indicates severe sleepiness

This version is adapted from Sleep Disorders for Dummies (2004) (82).

of waking blend into the states of sleep. Hypnosis, meditation, and drowsiness approach the borderline of waking, sometimes extending across that threshold into dreaming consciousness.

Notes

1. Poincaré, H. (1929) *The Foundations of Science*. New York: Science House, p. 388.
2. Benson, D. (1994) *The Neurology of Thinking*. New York: Oxford University Press, p. 223.
3. Shoben, E. J. (1961) Culture ego psychology, and an image of man. *American Journal of Psychotherapy* 15: 395–408.
4. Wolman, R. and Kozmova, M. (2006) Last night I had the strangest dream: Varieties of rational thought processes in dream reports. *Consciousness and Cognition*. doi: 10.1016/j.concog2006.09.009; Kozmova, M. (2012) Dreamers as agents making

strategizing efforts exemplify core aggregate of executive function in non-lucid dreaming. *International Journal of Dream Research* 5: 47–67.

5. Hartmann, E. (1998) *Dreams and Nightmares – The New Theory on the Origin and Meaning of Dreams*. New York: Plenum. Beyond his wiliness to analytically address nightmare, this important emphasis on the variable aspects of waking consciousness is a primary contribution that Ernest Hartmann has made to the field of dream science.

6. Anderson, J. (2005) *Cognitive Psychology and Its Implications* (6th ed.) New York: Worth, p. 443.

7. Catell, R. (1963) Theory of fluid and crystallized intelligence. *Journal of Educational Psychology* 54: 1–22.

8. James, W. (1890) *The Principles of Psychology*, Vols. 1 & 2. New York: Holt, pp. 403–404.

9. Posner, M. and Petersen, S. (1990) The attention system of the human brain. *Annual Review of Neuroscience* 13: 25–42; Fan, J. McCandliss, B., Sommer, T., Raz, A. and Posner, M. I. (2002) Testing the efficiency and independence of attentional networks. *Journal of Cognitive Neuroscience* 14: 340–347.

10. Hobson, J. A. (1999) *Consciousness*. New York: Scientific American Press; Steriade, M. (2006) Grouping of brain rhythms in corticothalamic systems. *Neuroscience* 137: 1086–1106.

11. Crick, F. (1994) *The Astonishing Hypothesis: The Scientific Search for the Soul*. New York: Charles Scribner's Sons; Blaha, S. (2000) *Cosmos and Consciousness*. New Hampshire: Price Hill; Hobson, J. A. (1999) *Consciousness*. New York: Scientific American Press.

12. Buzsaki, G. (2005) Theta rhythm of navigation: Link between path integration and landmark navigation, episodic and semantic memory. *Hippocampus* 15: 827–840.

13. Buzsaki, G. (2006) *Rhythms of the Brain*. Oxford: Oxford University Press.

14. Axmacher, N., Mormann, F., Fernandez, G., Elger, C. and Fell, J. (2006) Memory formation by neuronal synchronization. *Brain Research Review* 52: 170–182; Corsi-Cabrerra, M., Guevara, M. and Rio-Portilla, Y. (2008) Brain activity and temporal coupling related to eye movements during REM sleep: EEG and MEG results. *Brain Research* 1235: 82–91; Lisman, J. E. and Idiart, M. A. (1995) Storage of 7 + 2 short term memories in oscillatory subcycles. *Science* 267: 1512–1515.

15. Shiffrin, R. (1997) Attention, automatism, and consciousness, in *25th Symposium on Cognition: Scientific Approaches of Consciousness*, ed. J. Cohen and J. Schooler. Hillsdale, NJ: Erlbaum, pp. 49–64.

16. Pasher, H. (1998) *The Psychology of Attention*. Cambridge, MA: MIT Press.

17. Anderson, J. (2005) *Cognitive Psychology and Its Implications* (6th ed.). New York: Worth, p. 105.

18. Buckner, R., Andrews-Hanna, J. and Schacter, D. (2008) The brain's default network – Anatomy, function and relevance to disease. *Annals of the New York Academy of Science* 1124: 1–38; Raichle, M., MacLeod, A., Snyder, A., Powers, W, Gusnard, D., et al. (2001) A default mode of brain function. *Proceedings of the National Academy of Sciences of the United States of America* 98: 676–682.

19. Buckner, R. and Carroll, D. (2007) Self projection and the brain. *Trends in Cognitive Science* 11: 49–57; Gusner, D. and Raichle, M (2001) Searching for a baseline: Functional imaging and the resting human brain. *Nature Reviews. Neuroscience* 2: 685–694.

20. Christoff, K., Gordon, A., Smallwood, J., Smith, R. and Schooler, W. (2009) Experience sampling during fMRI reveals default network and executive system contributions to mind wandering. *Proceedings of the National Academy of Sciences of the United States of America* 1–6.

21. Starker, S. and Singer, J. (1975) Daydream patterns and self-awareness in psychiatric patients. *Journal of Nervous and Mental Disease* 161: 313–317.

22. Pagel, J. F. (2007) Obstructive sleep apnea and AD/HD – Associated quantitative EEG abnormalities, in *Sleep Apnea Syndrome: Research Focus*, ed. A. Lang. Hauppauge, NY: Nova, pp. 107–127.

23. Andrews-Hanna, J. (2011) The brain's default network and its adaptive role in internal mentation. *Neuroscientist* XX: 1–20; Daoust, A., Lusignan, F., Braun, C., Mottron, L. and Godbout, R. (2008) EEG correlates of emotions in dream narratives from typical young adults and individuals with autistic spectrum disorders. *Psychophysiology* 45: 299–308.

24. Byron, K. and Khazanchi, S. (2012) A meta-analytic investigation of state and trait anxiety to performance on figural and verbal creative tasks. *Personality and Social Psychology Bulletin* 37: 269–283.

25. Pagel, J. F. (1993) Modeling drug actions on electrophysiologic effects produced by EEG modulated potentials. *Human Psychopharmacology* 8: 211–216; Pagel, J. F. (1996) Pharmacologic alterations of sleep and dream: A clinical framework for utilizing the electrophysiological and sleep stage effects of psychoactive medications. *Human Psychopharmacology* 11: 217–223.

26. Pagel, J. F. (2012) The synchronous electrophysiology of conscious states. *Dreaming* 22: 173–191.

27. Raiche, M. (2006) Neuroscience. The brain's dark energy. *Science* 314: 1249–1250; Fox, M. and Raichle, M. (2007) Spontaneous fluctuations in brain activity observed with functional magnetic resonance imaging. *Nature Reviews. Neuroscience* 8: 700–711.

28. Andrews-Hanna, J. (2011) The brain's default network and its adaptive role in internal mentation. Fox, M. and Raichle, M. (2007) Spontaneous fluctuations in brain activity observed with functional magnetic resonance imaging; Finglehurts, A. and Finglehurts, A. (2011) Persistent operational synchrony within brain default-mode network and self processing operations in healthy subjects. *Brain and Cognition* 75: 79–90; Horovitz, S., Fukunaga, M., Zwart, J., Van Gelderen, P., Fulton, S., Balkin, T. and Duyn, J. (2008) Low frequency BOLD fluctuations during resting wakefulness and light sleep: A simultaneous EEG–fMRI study. *Human Brain Mapping* 26: 671–682.

29. Buzsaki, G. (2005); Steriade, M. (2006) Grouping of brain rhythms in corticothalamic systems. *Neuroscience* 137: 1086–1106.

30. Besserve, M., Jerbi, K., Laurent, F., Baillet, S., Martinerie, J. and Garnero, L. (2007) Classification methods for ongoing EEG and MEG signals. *Biological Research* 20: 415–437.

31. Kiebel, S., Garrido, M., Moran, R., Chen, C. and Friston, K. (2009) Dynamic causal modeling for EEG and EMG. *Human Brain Mapping* 30: 1866–1876.

32. Fox, M. and Raichle, M. (2007); Leopold, D., Murayama, Y. and Logothetis, N. (2003) Very slow activity fluctuations in monkey visual cortex: Implications for functional brain imaging. *Cerebral Cortex* 13: 423–433.

33. Domhoff, W. (2013) The neural substrate for dreaming: Is it a subsystem of the default network? *Consciousness and Cognition* 20: 1163–1174.

34. Silvera, J. (1971) Incubation: The effect of interruption timing and length on problem solution and quality of problem process. Unpublished doctoral dissertation, University of Oregon, reported by Anderson, pp. 272–274.

35. Smith, S. and Blakenship, S. (1991) Incubation and the persistence of fixation in problem solving. *American Journal of Psychology* 104: 61–87.

36. Ellamil, M., Dobson, C., Beerman, M. and Christoff, K. (2012) Evaluative and generative modes of thought during the creative process. *NeuroImage* 59: 1783–1794; Chrysikou, E. and Thompson-Schill, S. (2012) Dissociable brain states linked to common and creative object use. *Human Brain Mapping* 32: 665–675.

37. Giambra, L. (1979) Sex differences in daydreaming and related mental activity from the late teens to the early nineties. *International Journal of Aging and Human Development* 10: 1–34.

38. Scarry, E. (1999) *Dreaming by the Book*. Princeton, NJ: Princeton University Press, pp. 3–4.

39. States, B. (1997) *Seeing in the Dark*. New Haven, CT: Yale University Press, pp. 3, 4, 198.

40. Pagel, J. F. (2008) *The Limits of Dream – A Scientific Exploration of the Mind/Brain Interface*. Oxford: Academic Press/Elsevier.
41. Polgelt, B. (1981) Relations between rhythmic brain processes and psychomotor tempo. *Activitas Nervosa Supererior (Praha)* 23: 97–101.
42. Lakatos, P., Karmos, G., Mehta, A., Ulbert, I. and Schroeder, C. (2008) Entrainment of neuronal oscillations as a mechanism of attentional selection. *Science* 320: 110–113; Palva, J., Palva, S. and Kaila, K. (2005) Phase synchrony among neuronal oscillations in the human cortex. *Journal of Neuroscience* 24: 3962–3972.
43. Gruzelier, J. (2009) A theory of alpha/theta neurofeedback, creative performance enhancement, long distance functional connectivity and psychological integration. *Cognitive Processing* 10(S1): S101–S109.
44. Fink, A., Graf, B. and Neubauer, A. (2009) Brain correlates underlying creative thinking: EEG alpha activity in professional vs. novice dancers. *NeuroImage* 46: 845–862.
45. Fink, A., Grabner, R., Benedek, M., Reishofer, G., Hauswirth, V., Fally, M., et al. (2009) The creative brain: Investigation of brain activity during creative problem solving by means of EEG and fMRI. *Human Brain Mapping* 30: 734–748.
46. Greenberg, I. A. (1974) *Psychodrama: Theory and Therapy*. New York: Behavioral Publications.
47. Shapiro, F. and Forrest, M. (1997) *EMDR: Eye Movement Desensitization and Reprocessing*. New York: Basic Books. Gruzelier, J. (2009) A theory of alpha/theta neurofeedback, creative performance enhancement, long distance functional connectivity and psychological integration.
48. Society of Psychological Hypnosis (2004) Division 30 of the American Psychological Association, Statement of Executive Committee.
49. Agogino, A. (1965) The use of hypnotism as an ethnographic research technique. *Plains Anthropologist* 10: 31–36.
50. Krippner, S. and Kremer, J. (2010) Hypnotic-like procedures in indigenous shamanism and mediumship, in Hypnosis and Hypnotherapy, ed. D. Barrett, Vol. 1 – Neuroscience, Personality and Cultural Factors. Santa Barbara, CA: Praeger, pp. 104–106.
51. Kunzendorf, R. (2010) The hypnotic deactivation of self-conscious source monitoring, in Hypnosis and Hypnotherapy, ed. D. Barrett, Vol. 1 – Neuroscience, Personality and Cultural Factors. Santa Barbara, CA: Praeger, pp. 1–3.
52. Spiegel, H. and Spiegel, D. (2004) *Trance and Treatment: Clinical Uses of Hypnosis*. Washington, DC: American Psychiatric Publishing.
53. Barrett, D. (2010) Hypnotic dreams, in Hypnosis and Hypnotherapy, ed. D. Barrett, Vol. 2 – Application in Psychotherapy and Medicine. Santa Barbara, CA: Praeger, pp. 99–122.
54. Spiegel, D., White, M. and Waelde, L. (2010) Hypnosis, mindfulness meditation and brain imaging, in Hypnosis and Hypnotherapy, ed. D. Barrett, Vol. 1 – Neuroscience, Personality and Cultural Factors. Santa Barbara, CA: Praeger, pp. 38–52.
55. Katayama, H., Gianotti, L., Isotani, T., Faber, P., Sasada, K., Kinoshita, T. and Lehmann, D. (2007) Classes of multichannel EEG microstates in light and deep hypnotic conditions. *Brain Topography* 20: 7–14. Pascalis, V. (1999) Psychophysiological correlates of hypnosis and hypnotic susceptibility. *International Journal of Clinical and Experimental Hypnosis* 47: 117–143.
56. Katayama, H. et al. (2007) Classes of multichannel EEG microstates in light and deep hypnotic conditions.
57. Buckley, K. (2009) *Dreaming and the World's Religions*. New York: New York University Press.
58. Thera, N. (1998) *Buddhist Explorations of Consciousness and Time*. Boston, MA: Wisdom Publications, p. 121.

59. Flickstein, M. (2007) *The Meditators' Atlas – A Roadmap of the Inner World*. Boston, MA: Wisdom Publications, p. 89.

60. Spiegel, D., White, M. and Waelde, L. (2010) Hypnosis, mindfulness meditation and brain imaging, in *Hypnosis and Hypnotherapy*, ed. D. Barrett, Vol. 1 – Neuroscience, Personality and Cultural Factors. Santa Barbara, CA: Praeger, pp. 38–52.

61. Seeley, W., Menon, V., Schatzberg, A., Keller, J., Glover, G. H., Kenna, H., et al. (2007) Dissociable intrinsic connectivity networks for salience processing and executive control. *Journal of Neuroscience* 27: 2349–2356.

62. Fell, J., Axmacher, N. and Haupt, S. (2010) From alpha to gamma: Electrophysiological correlates of meditation-related states of consciousness. *Medical Hypotheses* 75: 218–224.

63. Lutz, A. Greischar, L., Rawlings, N., Richard, M. and Davidson, R. (2004) Long-term meditators self-induce high-amplitude gamma synchrony during mental practice. *Proceedings of the National Academy of Sciences of the United States of America* 101: 16369–16373; Huang, H. and Lo, P. (2009) EEG dynamics of experienced Zen meditation practitioners probed by complexity index and spectral measure. *Journal of Medical Engineering and Technology* 33: 314–321; Cahn, B., Delorme, A. and Polich, J. (2010) Occipital gamma activation during Vipassana meditation. *Cognitive Processing* 11: 39–56.

64. Vialatte, F., Bakardjian, H., Prasad, R. and Cichocki, A. (2009) EEG paroxysmal gamma waves during Bhramari Pranayama: A yoga breathing technique. *Consciousness and Cognition* 18: 977–988.

65. Lehmann, D., Faber, P., Achermann, P., Jeanmonod, D., Gianotti, L. and Pizzagalli, D. (2001) Brain sources of EEG gamma frequency during volitionally meditation-induced, altered states of consciousness, and experience of the self. *Psychiatry Research* 108: 111–121.

66. Huang, H. and Lo, P. (2009) EEG dynamics of experienced Zen meditation practitioners probed by complexity index and spectral measure; Chiesa, A. (2009) Zen meditation: An integration of current evidence. *Journal of Alternative and Complementary Medicine* 15: 585–592.

67. Chan, A., Han, Y. and Chueng, M. (2008) Electroencephalographic (EEG) measurements of mindfulness-based Trirchic body-pathway relaxation technique: A pilot study. *Applied Psychophysiology and Biofeedback* 33: 39–47.

68. Lagopoulos, J., Xu, J., Rasmussen, I., Vik, A., Malhi, G. and Eliassen, C., et al. (2009) Increased theta and alpha activity during nondirective meditation. *Journal of Alternative and Complementary Medicine* 15: 1187–1192; Cahn, B., Delorme, A. and Polich, J. (2010) Occipital gamma activation during Vipassana meditation. *Cognitive Processing* 11: 39–56.

69. Pagel, J. F. (October 2012) Personal report – interview with Xin Yuan, Vice Abbott – Jingza Monastery, China.

70. Rinpoche, T. W. (1998) *The Tibetan Yogas of Dream and Sleep*. Ithaca, NY: Snow Lion.

71. Buysse, D. J. (1991) Drugs affecting sleep sleepiness and performance, in *Sleep, Sleepiness, and Performance*, ed. T. M. Monk. Chichester: Wiley & Sons, pp. 4–31.

72. National Sleep Foundation (2000) *2000 Omnibus Sleep in America Poll*. Washington, DC: National Sleep Foundation.

73. Friedman, N. (2007) Determinates and measures of daytime sleepiness, in *Primary Care Sleep Medicine – A Practical Guide*, ed. J. F. Pagel and S. R. Pandi-Perumal. Totowa, NJ: Humana Press, pp. 61–82.

74. Horne, J. A. and Reyner, L. A. (1995) Sleep related vehicle accidents. *British Medical Journal* 310: 565–567. While the data are of concern, many have chosen to ignore the effects of sleepiness on driving while emphasizing the effects of more easily measurable and prosecutable drugs of abuse. In some ways this approach parallels the proposal that hallucinatory trance states rather than dreams are an explanation for the

ecstatic component of cave art. Both dreams and daytime sleepiness are "normal" factors that profoundly affect our waking functioning.

75. National Transportation Safety Board (1995) *Factors that Affect Fatigue in Heavy Truck Accidents*. Study NTSB/SS-95/01. Washington, DC: NTSB.

76. Pagel, J. F., Forister, N. and Kwiatkowski, C. (2007) Adolescent sleep disturbance and school performance: The confounding variable of socioeconomics. *Journal of Clinical Sleep Medicine* 3: 19–23.

77. Pagel, J. F. (2006) Medications that induce sleepiness, in *Sleep – A Comprehensive Handbook*, ed. T. Lee-Chiong. Hoboken, NJ: John Wiley and Sons, pp. 175–182.

78. Pivik, R. T. (1991) The several qualities of sleepiness, in *Psychophysiological Considerations in Sleep, Sleepiness and Performance*, ed. T. M. Monk. Chichester: John Wiley & Sons, pp. 3–38.

79. Bonnet, M. (2000) Sleep deprivation, in *Principles and Practice of Sleep Medicine* (3rd ed.), ed. M. Kryger, T. Roth and W. Dement. Philadelphia, PA: W.B. Saunders, p. 57.

80. Pagel J. F. Pandi-Perumal, S. R. (eds.) (2007) *Primary Care Sleep Disorders A Practical Guide*. New York: Humana Press.

81. Iber, C., Ancoli-Israel, S., Chesson, A. and Quan, S. (2007) *The AASM Manual for the Scoring of Sleep and Associated Events*. Westchester, IL: American Academy of Sleep Medicine.

82. Hirshkowitz, M. and Smith, P. (2004) *Sleep Disorders for Dummies*. Hoboken, NJ: Wiley. The Epworth Sleepiness Scale was originally developed in Australia by Dr. Murray Johns.

83. Anastassiou, C., Perin, R., Markam, H. and Koch, C. (2011) Emphatic coupling of cortical neurons. *Nature Neuroscience* 14: 217–223. Cash, S., Halgren, E., Dehghani, N., Rossetti, A., Thesen, T., Wang, C., et al. (2009) The human K-complex represents an isolated cortical down state. *Science* 324: 1084–1087; Steriade, M. (2006) Grouping of brain rhythms in corticothalamic systems.

84. Singer, W. (1999) Neuronal synchrony: A versatile code for the definition of relations? *Neuron* 24: 49–65; Salinas, E. and Sejnowski, T. (2001) Correlated neuronal activity and the flow of neural information. *Nature Reviews. Neuroscience* 2: 539–550; Cantero, J. and Atienza, M. (2005) The role of neural synchronization in the emergence of cognition across the wake–sleep cycle. *Reviews in the Neurosciences* 16: 69–83.

85. Pagel, J. F. (2005) Neurosignals – Incorporating CNS electrophysiology into cognitive process. *Behavior and Brain Science* 28: 75–76. Pagel, J. F. (2012) The synchronous electrophysiology of conscious states.

The Conscious Forms of Sleep

There is another world but it is in this one. **(Eluard, 1924) (1)**

Dreaming qualifies sleep states as conscious states. The perceptual and motor isolation of sleep prevents the description of this mental activity of dreaming until after an individual has awakened. This makes dreaming more difficult to approach, limiting the study of dreaming to the individual's dream report. Dreaming can be described objectively using reported recall frequency, and how dreaming affects waking behavior. However, the content of any dream is subjective, reflecting an individual's experience of the waking world. That subjectivity is best approached by example. In this chapter, in addition to objective descriptions, individual representative dreams are included as examples. In most cases, these are dreams experienced by the author of this text.

Sleep, like waking, is a varied state that can be divided into different stages. These stages of sleep have inherent differences in electrophysiology, neuroanatomy, and function. Reports of the mentation occurring in each sleep stage are also affected by the transition into wakefulness. Occurring in the isolation of sleep, all dreaming states are affected by this transition. As Graham Cairns-Smith notes, "the process of waking shows us each morning that there are no sharp distinctions between dreams and other forms of our conscious being" (2). We traverse across the diffuse border from the various states of dreaming sleep, to dreaming awake, and finally to the state of "dreaming aware" in which we integrate the conscious experiences of sleep with the structure of our surroundings (3).

The accessibility of dreaming to waking consciousness is based on the cognitive and electrophysiological distance of each sleep state from waking (4). At that transition, the reports of mental activity differ based on the sleep stage from which the individual is awakening. These reports, which we call dreams, provide insight into the phenomenological differences among the various forms of sleeping consciousness. Some dreaming states exist just across the borderline from waking. Individuals arouse easily from these states and often report detailed content of their experienced mentation. Some of these states can be difficult to differentiate from waking, whether based on behavior, electrophysiology, or neuroanatomical activation. Other dreaming states occur during states of sleep that are vastly different from waking. The dreams of deep sleep (stage 3) include content and behavior on arousal that are bizarre and markedly different from waking experience. This phenomenology is associated with patterns of brain electrical activity and neuroanatomical deactivation that are rarely, if ever, seen in waking.

Behaviorally, sleep is a state of reversible perceptual dissociation. This lack of perceptual input is its primary behavioral characteristic and the primary factor contrasting the sleeping from the waking states. However, perceptual isolation is not an absolute during sleep. Sounds, particularly the sound of your name, and noxious stimuli are sometimes incorporated into the content of dreams. Experimentally, perceptual integration occurs most often in stages of sleep such as rapid eye movement sleep (REMS), sleep onset (stage 1), and stage 2, that are physiologically closer to waking consciousness (5). Perceptual dissociation is at least a partial characteristic of the dream-like states such as lucid dreaming and sleep meditation that exist on the border between wakefulness and sleep. Characteristically in REMS, and at least partially in other sleep stages, a motor block exists preventing the potential actions or expressions of sleeping thoughts so that the physical acting out of a dreams does not usually occur.

Many of the sleep-associated states of consciousness are difficult to study because they cannot be consciously induced. These states reveal themselves only sporadically as non-consciously controlled parasomnias – unwanted and disruptive sleep-associated behaviors and dream-like experiences occurring most often after an abrupt waking from sleep. While some parasomnias such as limb movements and heart rate abnormalities are not usually associated with dreaming, many parasomnias include dream-like mentation. Hypnogogic hallucinations, sleep starts, and sleep paralysis are parasomnias occurring at sleep onset. Sleep talking, anxiety, and panic attacks are associated with stage 2 sleep. Nightmares, sleep paralysis, and REMS behavior disorder (RBD) occur most often in association with REMS (6). Other parasomnias occur out of deep sleep (stage 3). These parasomnias, described as "arousal disorders," include sleep terrors, sleep walking, and confusional arousals. All of the parasomnias occur more often in individuals with disorders such as sleep apnea and periodic limb movements that induce an increased number of arousals and awakenings from sleep. While difficult to study, parasomnias are interesting from the perspective of consciousness since their associated behaviors provide insight into the cognitive processes present in the underlying states of sleep.

Cognitively, the conscious states of sleep are more interesting and diverse than the conscious states of waking. In waking, we must address the perceptual analysis and integration of the exterior world, as well as the motor actions, learning activities, work, relationships, distraction, games, and entertainments of daily life. During waking cognition, states can have dream-like characteristics. However, such dream-like mind wandering occurs during intermissions in the concrete processes involved in the functional reality of focused waking. While dreaming may be as "thin as thought," the thought processes involved in dreaming are varied and complex (7). Dreaming is an aspect of reflective and tertiary consciousness. It is a compilation, integration, and critique of waking perceptions, senses, behaviors, and emotional experience. In some forms of dreaming, we stretch our capacities for visual imagery. In others, we confront extremes of emotional processing. In states such as lucidity, we can approach the limits of our capabilities for behavioral control. And then there are those states in which we might be able to apprehend the threshold of non-consciousness, and bring back to waking an impression of a farthest borderline of consciousness.

CONTROLLING DREAMING: LUCIDITY AND SLEEP MEDITATION

Lucid dreaming was initially described as "being awake in your dreams" and as "dreaming while being fully conscious that you are dreaming" (8). Currently, lucid dreaming is often described as the experience of achieving conscious awareness of dreaming while asleep (9). This description is phenomenologically diffuse since most dreams have lucid characteristics. Dreams are described from the dreamer's point of view (see Table 1.1 in Chapter 1). The dreamer is present, and at some level aware, during the mental activity of every dream.

Lucid dreaming has three primary characteristics: sleep state association, conscious control, and the capacity for volitional motor activity. The first requirement is that the individual be asleep. Much of the research into this state has focused on identifying and attempting to prove that lucid dreaming occurs during REMS (10). While lucid dreaming is generally considered as occurring in REMS, this postulate was developed during a period in which all dreaming was thought to occur during REMS. Even then, researchers noted that up to eighteen percent of lucid dreams occurred at sleep onset, a period during which REMS rarely occurs (11). Lucid dreaming can take place during stage 2 sleep and at sleep onset (stage 1). Despite this state's lack of isolation from the external environment, and questions about the sleep stage of association, most evidence indicates that lucid dreaming occurs during sleep (12).

The second characteristic of lucid dreaming is conscious control. During episodes of lucid dreaming, the dreamer can consciously control the action and content of the dream. In teaching lucidity, an individual is often instructed to imagine the visualization of a body part using waking imagery before initiating sleep. After that visualization is accomplished during dreaming, the individual is instructed to take active control of the dream character and the storyline of the dream. Therapeutically, this approach is sometimes used with apparent success to treat nightmare disorder and post-traumatic stress disorder (PTSD), clinical situations in which disturbing nightmares interfering with sleep result in waking distress. The patient is asked to write out or draw a version of his or her worst nightmare, and then rewrite that dream story to have a more positive or acceptable ending. Each night before sleep, the individual contemplates that revised imagery. In up to ninety percent of cases, this approach leads to a marked decline in nightmares and nightmare-related stress (13). Some individuals become experts at dream control and manipulation, achieving what has become the holy grail of lucidity work: the lucid dream within a dream.

The assertion of conscious control during dreaming can have less beneficial effects. I rarely have nightmares, and when I have them, they often occur in a dream version of Kauai, the Hawaiian island where I used to live. In my dream:

> I'm crossing a taro field knee deep in muddy water, following a straight path cut between plants, and followed in line by my wife and daughter. The sky is blue. The dirt is red. The ocean stretches to the horizon in the distance. In the path a large snake swims sinuously directly towards us. There is no place to go, nowhere to run in the deep mud. There is also no way to protect my wife or daughter. The snake approaches, undulating back and forth, its mouth opening to reveal giant fangs dripping with venom. I change the nightmare [there is some question as to whether I did

this from within the dream or during a mini-arousal before continuing the dream], and the snake becomes a brightly colored pool blow-up toy snake, inflating larger and larger – something to be laughed at. I push it aside and we continue across the wet field to safety on the other side.

I still vividly remember the dream the following morning and experience no distress or discomfort. Months later, in reflection, I realize that in exerting control during the dream, I had mentally turned and run from the snake and its associations. My dream had offered me a message and an understanding of past conflict returning once again to bedevil me, my wife, and especially my child. Parodying the nightmare snake as a harmless inflatable toy, while an option that diminished my distress, was not particularly helpful when used with lucid control during dreaming to avoid distress and discomfort. This dream, this nightmare, was providing a possibility for insight and knowledge, an insight that with my abilities at lucid control, I chose to avoid and not use.

The third, and what is experimentally the most interesting characteristic of lucid dreaming is the dreamer's capacity for volitional motor activity during what appears to be sleep. During lucid dreaming, some individuals can move or fix their gaze, using this change to signal an external observer (14). Others can push buttons during episodes of lucidity. This capability can be taught, but it is evidently difficult to learn. Because of the rarity of such subjects, research studies are sometimes conducted and based on the analysis of only one individual, and that individual is sometimes the author of the study. Methodologically, this is an approach that makes it very likely that study results will reflect the perspective and bias of the subject/author of the study (15). Yet, this volitional signaling capability has allowed researchers to analyze lucid dreaming with external real-time scanning modalities. The initial hope for this approach was that scanning would display areas of the central nervous system (CNS) that are active during REMS dreaming. However, the results of such studies indicate that lucid dreaming with signaling is not associated with the brain areas known to control either REMS or the motor paralysis associated with REMS (16). During lucid dreaming, there are multiple activated sites of the CNS that are normally deactivated during REMS. The activated CNS sites are among those that are known to be involved in working memory and the analysis of visual perception (17). These functional magnetic resonance imaging (fMRI) data have been interpreted by the authors of these studies to suggest that lucid dreaming is a vastly different state from other forms of dreaming, involving memory and visual systems that are not part of any other form of dreaming or sleep (18).

Bursts of alpha and gamma electroencephalograph (EEG) frequency occur during lucid dreaming. These same EEG patterns are seen during focused and creative waking, default waking, meditation, and lucid dreaming (see Chapter 5). All are teachable states of focused awareness under

TABLE 6.1   Comparative Behavioral Characteristics for Sleeping-Associated States

	Perceptual Isolation	Thought-Attention	Memory Access	Mentation (Dream) Recall	Volitional Control	Teachability
Hypnogogic states	Moderate	Disassociated	High	80%	Low – variable	Moderate
Stage 2 dreaming	Moderate	Continuity	High	40%	Low	Low
REMS dreaming	High	Focused	High	80%	Low	Low
Deep sleep (stage 3) dreaming	High	Unfocused	Low	40%	Low	None
Lucid dreaming	Low	Focused	High	80%	High	Moderate
Sleep meditation	High	Focused	High	High	High	High

Note: REMS = rapid eye movement sleep.

conscious control, with similarities in reported effects on cognitive processing, attention, and recall (see Table 5.2 in Chapter 5, and Table 6.1) (19). Electrophysiologically, the signaling used to experimentally mark lucidity is almost always associated with an episode of arousal from sleep. Based on this EEG evidence, lucid dreaming may be best viewed as a transitional state between sleep and waking, with electrophysiological characteristics consistent with the research criteria used in establishing the state (volitional eye movements and/or button pushing). Such signaling, required to establish the state, leads to arousals from sleep (20). Lucid dreaming that includes such signaling is a state in which the sleeper exerts conscious thought and motor control, and with this signaling, the lucid dreamer is actually making a transition to waking perception and motor control.

The activation of working memory and visual systems seen on scanning fMRI and quantitative electroencephalography (QEEG) is most likely occurring secondary to activation of the parts of the CNS required for the lucid dreamer to be able to signal the watching researchers. These laboratory studies of lucid dreaming that require visual or motor signaling are describing conscious sleep offset – the opposite of the turning off of consciousness required for sleep onset. The sleeper is consciously reinstituting components of working memory and the visual activity systems associated with wakefulness in order to signal to the researchers that he or she is conscious (21).

Lucidity and the capacity for the conscious awareness of dreaming within a dream can be a remarkable experience. However, these findings argue that the state of dream lucidity may best be experienced without attempting to intrude into waking:

> Let us imagine the dreamer: in the midst of the illusion of the dream world and without dreaming it, he calls out to himself: "It is a dream, I will dream on." What must we infer? That he experiences a deep inner joy in dream contemplation; on the other hand, to be at all able to dream with this inner joy of contemplation, he must have completely lost sight of waking reality and its ominous obtrusiveness. (Nietzsche, 1968) (22)

Meditation initiated before sleep onset and continuing into sleep is an approach used in some Tibetan Buddhist meditative traditions (23). Sleep-associated meditation is probably a form of lucid dreaming, since the dreamer/meditator is exerting volitional control of dream content (24). Electrophysiologically, both focused meditation and lucid dreaming are associated with high levels of alpha and beta/gamma frequencies. This suggests that a correlation may exist between these behaviorally lucid states. Lucid dreaming may be best viewed as a trainable meditative-like state that, unlike other forms of waking-based meditation, is developed and elaborated while in the state of sleep.

In lucid dreaming and sleep meditation, we are observing a limit for the human capacity for cognitive control of thought. While in some traditions, perceptual isolation is a primary goal of waking meditation, in lucidity and sleep meditation the dreamer attempts to approach the cognitive borderline of waking perception and motor activity from within the isolation of sleep. The other conscious states of dreaming are on the other side of that line, outside the limits of perceptual experience and conscious control. They take place fully within sleep.

SLEEP-ONSET DREAMS AND HALLUCINATIONS

> One evening, therefore, before I fell asleep, I perceived, so clearly articulated that it was impossible to change a word, but nonetheless removed from the sound of any voice, a rather strange phrase which came to me without any apparent relationship to the events in which, my consciousness agrees, I was then involved, a phrase which seemed to me insistent, a phrase, if I may be so bold, which was knocking at the window. I took cursory note of it and prepared to move on when its organic character caught my attention. Actually, this phrase astonished me: unfortunately I cannot remember it exactly, but it was something like: "There is a man cut in two by the window," but there could be no question of ambiguity, accompanied as it was by the faint visual image. (Breton, 1924) (25)

If awakened just after initiating sleep, more than eighty percent of us will report dreaming. This high level of recall is the same frequency of dream recall reported from REMS (26). Yet, these sleep-onset dreams have received limited study, since for the past fifty years dreaming theoretically was occurring only during REMS. Most of what we know of sleep-onset dreaming comes from the study of hypnogogic (sleep-onset) parasomnias. Hypnogogic hallucinations occur in otherwise normal people at the onset of sleep (see the quote from André Breton, above). These are true hallucinations, most often auditory but frequently visual, and seemingly real to the dreamer. These experiences can be quite bizarre and frightening, ranging from the sounds of a dog barking, a baby crying, or an alarm ringing that wakes the dreamer, to the extreme experience of suffocation at the hands of a succubus. Sleep paralysis, a parasomnia more commonly associated with REMS, can also occur at sleep onset. During these experiences, the individual is paralyzed and unable to voluntarily move. Sleep paralysis often is associated with negative, disturbing dream content. These experiences can be terribly frightening. It is believed that Fusilli's famous painting "The Nightmare" is a rendition of an experience of sleep paralysis. While both hypnogogic hallucinations and sleep paralysis occur commonly in individuals with the neurological disease of narcolepsy, these parasomnias are quite common in the general population. More than half of Americans report having experienced one or the other (27).

In the next chapter we will look more closely at the nightmares associated with PTSD. PTSD nightmares often include repetitive concrete re-experiencing of the physical and/or psychological trauma that led to the individual's diagnosis of PTSD. When these nightmares occur at sleep onset, they can lead to insomnia, hyperarousal, and intense anxiety.

Artists and writers have sometimes induced hypnogogic dreaming and used the resultant dream experiences in their work. Both Salvador Dali and John Keats are known to have used this technique. They would attempt to fall asleep while sitting in a chair and holding a coin between thumb and index finger. When they fell asleep, their hands would relax, and the coin would fall into a dish set beside the chair, startling them awake. There are those who insist that sleep-onset dreams are less bizarre than REMS dreams (28); the strange and surrealistic images that Dali derived using this technique argue otherwise.

Sleep-onset dreams are different from the dreams of other stages of sleep. They are short in duration. They often include intense visual imagery and only limited content or story. They are often reality based, and are sometimes associated with intense distress and anxiety. They can be exceedingly bizarre.

Assessment of "bizarreness" has become a matter of contention in dream science. In the 1950s, Hall and Van de Castle designed an analytical scale used for recording and statistically comparing dream content. This classic and well-validated scale does not address bizarreness, a

TABLE 6.2   Bizarreness Scales

Dream Content Bizarreness: Hobson (Ho) Scale	
1: Discontinuities	Change of identity, time, or place
2: Incongruities	Mismatching features
3: Uncertainties	Explicit vagueness
Frequency of bizarre dream attributes (Ho scale)	
A1: Plot, character, or action discontinuities	44%
B2: Thought incongruities	16%
B3: Thought uncertainties	29%
Dream Content Bizarreness: Hunt (Hu) Scale	
Hallucination: perceptual transformations	Visual – somatic – auditory
Clouding/confusion	Abrupt changes in scene
	Gaps in dream
	Disorganized memory and thought
	Difficulties in recall after waking
General	Uncanny emotion
	Mythic thought
	Bizarre personification

characteristic that has become important to dream scientists attempting to support a special relationship between REMS and dreaming. Recently, the Hobson (Ho) scale was developed, designed in a way that makes REMS dreams appear the most bizarre (consistent with the theoretical constructs of the scale designer). This scale rates bizarreness based on the storyline content of the dream. Dreams that include incongruous, uncertain, and discontinuous wording and story are rated as the most bizarre (Table 6.2). Compared to dreams from the other sleep stages, REMS dreams tend to be longer, and include more words and a more developed storyline. Studies using this bizarreness scale consistently demonstrate that REMS dreams are the most bizarre. Based on the Hobson scale, sleep-onset dreams, owing their short and limited storyline, are less bizarre than REMS dreams. This is a consistent finding, despite the apparently "bizarre" hallucinations, extreme emotional distress, and intense disassociation from reality associated with sleep-onset dreaming. An alternative bizarreness scale developed by Harry Hunt is included for comparison (Table 6.2) (29). Sleep-onset dreams score very differently when rated on this scale. Ratings of bizarreness are an excellent example of how experimental methodology can be manipulated by researchers to achieve their desired results.

Dreams commonly occur at sleep onset. Most of us can describe our own sleep-onset dreams. Mine typically occur during those early mornings when I get up early to sit in meditation. While sitting and staring at the wall, I am prone to microsleeps associated with intense short and visual dream experiences: a beautiful young brunette holding a microphone, wearing a red dress, leaning on a bar-stool, alone at a bar upholstered with red simulated leather; a corner of yellow-and-white stitched drapery frozen motionless in a window that opens to the midday sky; a partial visual field image, the right upper quadrant, of a wide angled view of ocean rollers, white on blue, extending impossibly in each direction to the curve of the Earth. Each hypnogogic experience has common characteristics: they are short, visually and emotionally intense, reality based, relatively bizarre, and associated with limited story. Each occurs with the initiation of perceptual dissociation at the onset of sleep, forming a snapshot of non-perceptual visual consciousness present at sleep onset: consciousness without perception, without associated content and memories, without control, yet with intense emotion, visual intensity, and detailed recall. This is sleep-onset consciousness. Evidently, within the structure of meditative practice, such events are quite common. One teacher was quoted, when approached by a student excited by an intense visual image that had occurred during meditation, as dogmatically stating, "What is the value of such a blank insight? Don't dream!" (30).

The formal characteristics of sleep-onset dreams are summarized in Table 6.3. Comparative bizarreness per scale is ranked, as well as characteristic dream state-associated thought processes as based on work by Wolman and Kozmova (see Table 5.1 in Chapter 5) (31).

THE "SPECIAL" RELATIONSHIP OF RAPID EYE MOVEMENT SLEEP WITH DREAMING

All mammals, almost all monotremes, and many birds have the electrophysiological state of REMS. Most humans, even those with extensive neurological damage, have episodes of REMS that continue to occur into extreme old age. Neurophysiologists are fascinated by the state, and since animal models can be utilized in research, the neuroanatomy and neurochemistry of REMS are better described than those of any of the other states of sleeping consciousness.

As noted in the Preface, for much of the past fifty years it has been presumed that all dreaming occurs during REMS. Because of this presumption, few dream scientists considered that REMS dreams might be different

TABLE 6.3   Sleep-Onset Consciousness: Formal Characteristics, Bizarreness, and Thought Processing

Formal Characteristics	Bizarreness		Rational Thought Processing
	Hobson Scale	Hunt Scale	
Primarily visual hallucinations perceived as potentially real	Discontinuities – high	Hallucinations – high	Analytical – low
Coherent dream stories	Incongruities – low	Clouding/ confusion – high	Perceptual – high
High recall	Uncertainties – low	General – high (emotion, bizarre personification)	Memory and time awareness – high
Potential lucidity			Affective – high
Impression of falling (sleep starts)			Executive – low
Intense anxiety (PTSD, sleep paralysis)			Subjective – high
Recurrent (reality based) PTSD			Intuitive/ projective – low
			Operational – low

Note: PTSD = post-traumatic stress disorder.

from the dreams of other sleep states. In the past few years, many of the dream science theorists, forced to admit that dreams occurred without REMS and REMS without dreaming, have assumed the perspective that while dreams may occur outside REMS, it is only the state of REMS that has a "special" relationship with dreaming. They argue that REMS dreams are the classic big dreams of psychoanalysis, creative discovery, and religious and ecstatic insight. These theorists insist that REMS dreams are the only "real" dreams. They assert that REMS dreams are the most emotional, are the most bizarre, are associated with the highest recall of any sleep state, are the only dream state with lucidity, and are the only sleep mentation with "dream-like" content (32). Some of these arguments are spurious. We recall dreams from sleep onset at the same frequency that we report them on awakening from REMS. The dreams of sleep onset and deep sleep are also associated with intense emotion. Bizarreness is higher in REMS only when a biased assessment scale is utilized (see Tables 6.2 and 6.3). Lucidity also occurs in stage 2 and with sleep offset and onset (33). Dream content studies designed to eliminate transfer effects and researcher bias have indicated that the content of REMS dreaming may not be significantly different from that of non-REMS dreams (34).

But REMS dreams are different from other dreams. REM dream reports are longer and include more words than dream reports from other sleep stages (35). A disproportionate number of the long dreams reported in both literature and psychoanalysis (Descartes' dreams, see Chapter 3; Freud's dream of Irma's injection, see Chapter 2) are probably dreams that occurred during REMS. REMS dreams, owing to their length, complexity, and organizational requirements for presentation, are usually reported on awakening as narrative stories (see Chapter 8). Their content, just like the content of dreams from other stages of sleep, is most often based on waking experience. They most resemble the narrative genre that we call "soap operas" (36).

Many parasomnias occur in association with REMS – nightmares are the most common of REMS parasomnias. Nightmares are disturbing mental experiences occurring during sleep that often result in awakening. Except in individuals with PTSD, nightmares occur almost exclusively during REMS. Typically, a nightmare begins as a seemingly real and coherent dream sequence that becomes increasingly more disturbing as it unfolds. Nightmares often include the negative emotions of anxiety, fear, or terror, as well as, anger, rage, embarrassment, and disgust. Content most often focuses on imminent physical danger (e.g., threat of attack, falling, injury, death), but may also include aggression toward others, potential personal failures, and other distressing themes such as suffocation (37). Nightmare dream imagery is often experienced as an unmitigated perception of external reality (nightmares seem very real) (38). Nightmares occur at their highest frequency in young females, and are reported least by older males (39). Some therapists have extended a belief in the association of nightmares with trauma into their "therapeutic" practice, searching for memories of unremembered family and childhood trauma in any individual presenting a history of recurrent nightmares. Since nightmares occur at the highest frequency in young women, this "therapeutic" approach often becomes a search for otherwise unremembered childhood sexual abuse. This search for "false memories" has sometimes led to "therapy-induced" personal and family disarray. It is clear that even in young women, recurrent nightmares often occur without any history of trauma. Five percent of the general population report nightmares to be a problem. Among insomniacs, nightmares are reported at even higher frequencies (40). There are typical personality patterns in the individuals who experience recurrent nightmares. These include fantasy proneness, psychological absorption, dysphoric daydreaming, and "thin" boundaries (the tendency to see the world in shades of gray rather than in black and white) (41). Individuals with recurrent nightmares are likely to have a creative or artistic focus in their daily lives. Sometimes they successfully utilize their nightmares in creative careers in writing, acting, and film making (42).

For those who suffer from nightmares, the associated distress, rather than the frequency of occurrence, is the complaint most likely to lead them to seek help. When an individual experiences significant psychological or physical trauma, recurrent nightmares can mark the functional failure of parts of the CNS involved in emotional processing. PTSD, the major dreaming-associated diagnosis, will be approached from this perspective in Chapter 7. The diagnostic and coding manuals for psychiatry and sleep medicine have historically confounded PTSD nightmares with the nightmares occurring in individuals without a history of trauma. It was only recently, in 2005, that a new diagnosis, nightmare disorder, was added to the categorization of sleep-associated diagnoses. Now this alternative diagnosis gives therapists and physicians the option of treating individuals with recurrent nightmares that disrupt their sleep without making a diagnosis of PTSD and searching for an often non-existent history of trauma (43).

Beyond nightmares, there are other REMS parasomnias that frequently include dream content. Sleep paralysis occurs most often during REMS. This parasomnia often includes negative and frightening content, developed in great detail, and associated with distress that extends into awakening. The dreams of REMS behavior disorder (RBD) are often physically aggressive interactions in which the dreamer is attacked by unfamiliar people or animals. Individuals with RBD sometimes exhibit complex and violent motor behaviors associated with their dream mentation. Injuries to the sleeper or bed partner are the most common symptom that leads them to seek medical attention. On video-polysomnographic monitoring, most of these individuals demonstrate increased muscle activity during REMS, a state that is otherwise defined by the flaccidity of motor atonia. This motor block – which keeps us from moving during dreaming – fails, and individuals with RBD apparently act out their dreams. The parasomnias of RBD sometimes occur outside REMS. In about half of affected individuals, RBD may herald the eventual onset of a neurodegenerative CNS illness, most often Parkinson's disease (44).

The author of this book, a late middle-aged male, is the appropriate gender and age for RBD onset. During the writing of this chapter:

> ... he dreams that he is writing on his computer while sitting in a rocking chair at a mountain log cabin. He is sitting at the front of the cabin's wrap-around porch, his legs propped up on the head and antlers of a dead elk that has been moved onto the porch to keep it away from bears. He looks up from his writing and into the forest, and sees a sow and two cubs standing aggressively just a few feet away. He jumps up, computer in hand, and runs around the corner of the cabin in his attempt to get away from the bears. He looks back the way that he had come, when something hits his elbow. He reacts as if touched by the bear, and his elbow comes down on the pillow, just missing his wife who was reaching out to quiet him.

Synchronicity or suggestion? Most likely a comment on the continuity of life's association with dreaming. Apparently, the study of parasomnias has its own risks for the writer and his mate. Dreaming bizarreness

TABLE 6.4   Rapid Eye Movement Sleep Consciousness: Formal Characteristics, Bizarreness, and Thought Processing

Formal Characteristics	Bizarreness		Rational Thought Processing
	Hobson Scale	Hunt Scale	
Coherent dream sequences	Discontinuities – high	Hallucinations – low	Analytical – high
Detailed plot, character, or actions	Incongruities – high	Clouding/ confusion – low	Perceptual – high
	Uncertainties – high	General – high (uncanny emotion)	Memory and time awareness – high
High recall			Affective – high
Increased length of report			Executive – high
Disturbing, intense emotions (nightmares and sleep paralysis)			Subjective – high
Potential lucidity			Intuitive/ projective – high
Recurrent (reality-based) PTSD			Operational – moderate
Comparative reality-based memory			

Note: PTSD = post-traumatic stress disorder.

(both scales) and thought processing characteristic of REMS-associated consciousness are summarized in Table 6.4.

ANXIETY (STAGE 2) DREAMING

We spend most of our sleep in stage 2, a stage that can be viewed as the junkyard of sleep. Stage 2 includes all of our sleep that is not sleep onset, REMS, or deep sleep. Electrophysiologically, stage 2 is defined by the presence of sleep spindles (episodic 15 Hz inclusions) and K-complexes (high-amplitude spikes/waves). Dream recall frequency is generally low, in the range of forty percent, and varies across the night. Since stage 2 includes the majority of sleep, dream researchers in attempting to emphasize the association between REMS and dreaming have extended this low recall percentage to all of non-REMS (45). During the "REMS equals dreaming" era, the mentation reported from other stages of sleep that was potentially non-REMS dreaming was described as follows: fragmentary brief experiences with thought-like content and disconnected visual imagery in which the dreamer is emotionally uninvolved. These non-REMS dreams were proposed not to be "genuine" dreams. It was suggested that they should be viewed as memory remnants from preceding REMS phases or as thoughts that developed after waking (46). It was suggested that this "degraded" mentation associated with the non-REMS stages was either "cognitive activity" without characteristics of actual dreaming or the result of "covert" REMS processing extending into non-REMS. In this era dreaming was the marker for REMS. Dreams developed in one sleep stage can continue throughout the night into later REMS and non-REMS periods of dreaming. Since dreaming was being reported from all sleep stages, it was proposed that REMS was "covertly" occurring throughout all of the sleep in which dreaming was present (47).

The content of stage 2 dreams differs from that of other dreams. There are fewer elements, a rather fragmented narrative structure, predominantly static impressions, and less emotional self-participatory involvement. When dreaming is defined as "imagery that consists of sensory hallucinations, emotions, story-like or dramatic progressions, and bizarreness ...," stage 2 dreaming, with few of these characteristics, could be considered not to be "genuine" dreaming (48). It is unclear whether this perceived difference in dream content is artifactual or real. These content data are based on studies that have not controlled for transference, researcher bias, or known differences in report length. One small study using methodology that controlled for these variables found that differences between REMS and non-REMS dreaming in emotional content disappeared, while differences in thought content persisted (49). While dream scientists have chosen to emphasize the differences there are more phenomenological similarities than there are differences among dreams based on their sleep stages of origin.

There have been few studies of either the electrophysiology or the neuroanatomy associated with stage 2. It is a diffuse and somewhat uninteresting state compared to sleep onset, REMS, or deep sleep. The dream-like parasomnias associated with stage 2 include sleep talking, anxiety dreams, and sleep panic attacks. These parasomnias share the characteristics of anxiousness and day-reflective content. The day-reflective characteristic of sleep talking can lead to a clinical presentation based on the repeating of topical information (e.g., calling out the name of a conjugal associate) that upsets the sleeping partner. Stage 2 dreams may extend throughout the night. They are rarely profound or particularly interesting. My personal favorite: I dreamed of a number – something that I rarely do – and I somehow thought that this number must be particularly important since it had come to me in a dream. I spent much of the night trying to remember that number. Each time I woke it would still be there, but I was very tired, and unable to summon the energy to get up and write it down. In the morning, I woke excited that I had remembered

TABLE 6.5   Stage 2 Consciousness: Formal Characteristics, Bizarreness, and Thought Processing

Formal Characteristics	Bizarreness		Rational Thought Processing
	Hobson Scale	Hunt Scale	
Day reflective content	Discontinuities – low	Hallucinations – low	Analytical – low
Associated anxiety and panic	Incongruities – low	Clouding/ confusion – low	Perceptual – moderate (perceptual integration associated with K-complexes)
Variable recall across night	Uncertainties – low	General – low	Memory and time awareness – low
			Affective – low
			Executive – low
			Subjective – low
			Intuitive/projective – low
			Operational – moderate (potential for arousal)

the number all through the night. As I became more fully alert, I was less excited: it was my own phone number.

The formal characteristics, bizarreness, and thought processing of stage 2 dreams and parasomnias are summarized in Table 6.5.

THE DREAMS OF DREAMLESS SLEEP

At the very threshold of consciousness there is the state of "dreamless sleep." Descartes was among the first to suggest that dreamless sleep is a state of non-consciousness. He suggested that sleep without dreaming could not exist. Since "thinking" defined consciousness, a non-thinking, non-dreaming state could be neither conscious nor alive (50). As discussed in Chapter 1, the ability to dream may mark the differentiation of our *Homo sapiens* ancestors from other non-dreaming proto-humans. Yet there are individuals functioning normally in our modern society who never recall their dreams (see Chapter 1).

There are a variety of reasons that individuals do not recall dreams. Some are structural, based on individual characteristics, personality, and pathology:

- There are rare individuals who have no dream recall (51).
- Major CNS injury in the basilar area of the frontal cortex can eliminate dreaming (52).

- Disordered sleep secondary to insomnia, obstructive sleep apnea, and depression can lead to a decline in recall (53).
- Some personality types (e.g., those with thick boundaries) have diminished dream recall (54).

Other reasons for a lack of recall are based on variables known to affect the process of recall. Time since dreaming, as well as distraction after waking, are associated with a decline in dream recall (55). Dream content characteristics, including salience and emotion, can increase our tendency to recall a particular dream (56). A lack of interest in dreaming, particularly in individuals without creative outlets, is also associated with diminished recall (57).

When we wake individuals in the sleep laboratory during sleep states with high recall propensity such as REMS and sleep onset, twenty percent of awakenings will be dreamless. In stage 3 (deep sleep) and stage 2, dream recall is reported from only forty percent of awakenings. In other words, up to one-third of sleep in otherwise normal individuals may be "dreamless." This lack of recall is based in part on the characteristics of the associated sleep state. As noted in the introduction to this chapter, some dream recall is probably lost during the required transition to wakefulness. The instability of the state transition from sleep to waking is likely to account for an overall twenty percent of reported loss of dream recall, since that is the portion of dream recall typically lost even in the high-recall states of sleep onset and REMS. The further diminishment of recall from stage 2 may be due to low dream salience secondary to the low emotionality, uninteresting content, and anxiety-based characteristics of dreaming in that state.

Deep sleep (stage 3) differs markedly from the other stages of sleep. Behaviorally, it is the most distant from waking and is the sleep stage least disturbed by external sounds or noxious stimuli. The reduced recall from this stage is due in part to its difference and distance from waking (58). The dream content of deep sleep is most often reported as an emotion, a visual color, or perhaps an awareness that dreaming has occurred. Behaviorally, we would understand very little about the mental activity of deep sleep if it were not for a series of remarkable parasomnias that occur on arousal from this state. These "arousal disorders" include sleep terrors, sleepwalking, and confusional arousals. These are common parasomnias, occurring in at least four percent of children. They often diminish or disappear with the onset of adolescence (59). In adulthood, they can be associated with psychiatric illness, sleep apnea, and RBD. Parasomnias occur rarely and spontaneously in the sleep laboratory, making them difficult to study. However, the arousal disorders are interesting from the perspective of consciousness since they tell us about the cognitive activity occurring in an otherwise unapproachable state.

BOX 6.1

SPINNING WHEELS - A DREAM OF DEEP SLEEP CONSCIOUSNESS

... darkness rumbling gray-wet, rock wheels set moving together in circuit. Green moss covering wheels that slowly roll together in a room deep underground. No matter how hard I tried to hold a tight focus, I would always falter and in that flicker of inattention, the wheels would begin to totter and lose their synchrony, their easy rolling. The rock wheels would crash into one another in eerie silence, smashing and shattering chamber walls, breaking out of darkness, sending me screaming out of sleep.

The associated automatic motor behavior occurring during these arousal disorders can be remarkably complex. Sleepwalkers can open doors, go on nocturnal wanders, and attempt complicated behaviors without the subject's conscious waking knowledge. Sleepwalkers are the inspiration for the zombies of literature and film. The emotional outbursts of night terrors include a fright-ridden cry, sweats, flushing, and apparent abject terror. Confusional arousals typically include disorientation, slow speech, diminished mentation, and inappropriate behaviors that can be vigorous, highly resistive, and violent. These states can last from a few minutes to several hours (60). The arousal disorders are the most bizarre and emotional of dream-like parasomnias, frightening and disturbing to the observer, who is often an upset and disturbed parent. The intense and frightening recurrent dream I had when eight years of age, of spinning, underground wheels with its scream on awakening, meets the diagnostic criteria for a night terror. For reference, it repeated here in Box 6.1. The formal characteristics of deep-sleep dreaming as well as characteristic thought and bizarreness are specified in Table 6.6.

Electrophysiologically, deep sleep is defined by the overwhelming presence of the delta (1 Hz) frequency, a rhythm rarely seen in other stages of sleep and almost never observed during waking. In some individuals, both K-complexes and sleep spindles can occur in stage 3. In youth, growth hormone is released during deep sleep. Beyond middle age, deep sleep is rarely seen, and what remains of the delta frequency declines in amplitude. This unique stage of sleep is physiologically important. An individual deprived of sleep, then finally allowed to sleep, will have sleep rebound that consists predominantly of stage 3. REMS rebound can occur, but it happens later and with less intensity, after there has been at least a partial recovery of the deep sleep that was lost (61).

TABLE 6.6   Deep Sleep Consciousness: Formal Characteristics, Bizarreness, and Thought Processing

Somnambulism (sleepwalking)	Autonomic and inappropriate behaviors, frantic attempts to escape a perceived threat, fragmentary recall
Confusional arousals	Mental confusion and disorientation on arousal from sleep, inappropriate and violent behaviors
Night terrors	Incoherent vocalizations, extreme fear, intense autonomic discharge, confusion and disorientation, fragmentary recall

Formal Characteristics	Bizarreness		Thought Processing
	Hobson Scale	Hunt Scale	
Mental confusion and disorientation on arousal	Discontinuities – low	Hallucinations – low	Analytical – low
Fragmentary recall	Incongruities – low	Clouding/confusion – high (low recall)	Analytical – low
Autonomic and inappropriate behaviors	Uncertainties – high (explicit vagueness)	General – high (emotion, bizarre personification)	Perceptual – low
Frantic attempts to escape a perceived threat			Memory and time awareness – low
			Affective – high
			Executive – low
			Subjective – low
			Intuitive/projective – low
			Operational – low

Benzodiazepines (e.g., Valium) affect the frequency and amplitude of the delta frequency associated with deep sleep. The effect can persist for days after ingestion. One of the benzodiazepines, clonazepam, is the drug most commonly used to treat the arousal disorders. Most antidepressants suppress deep sleep, and can be alternative medications for treatment of the arousal disorders. Gamma-hydroxybutyrate (one of the so-called date-rape drugs) and opiates can lead to an increase in delta frequency and an increased occurrence of arousal parasomnias (see Table 5.3 in Chapter 5) (62).

During deep sleep frontal cortex activity drops to a very low level. This is the one conscious state during which the default network is minimally active and apparently decoupled, potentially contributing to the suspension of consciousness during deep sleep (63). In deep sleep, we experience the sleep that is most likely to be "dreamless." The dreams of deep sleep that we do remember differ markedly from waking consciousness. Based on the deep-sleep parasomnias, these dreams can include minimal conscious control, automatic "zombie-like" behavior, screaming, extreme fright and terror, and diffuse, fragmentary recall of content. Like the "spinning-wheel" dream from my childhood, the content of such dreams, when brought to waking, can provide an understanding of mental activity occurring in the interior world of deepest sleep. Phenomenologically, the dream content associated with the arousal parasomnias is as close as we can come to describing the experience of this border of consciousness.

Some have used their skills at meditation in an attempt to empirically study borders of consciousness. These adepts describe states of perceptual disassociation in which thoughts do not occur. Electrophysiological and scanning studies of such experienced meditators indicate that these capacities are mirrored by physiological changes in brain activity (see Chapter 5). The recorded changes in brain activity include changes in theta, alpha, and gamma frequency activity, activity that is present in the high-recall dream states of REMS and sleep onset. Meditative changes in brain activity do not include changes in delta, the (1 Hz) frequency that defines deep sleep. This suggests that even the most skilled meditators have difficulty approaching this distant border of consciousness.

Deep-sleep consciousness is something far different from that which we normally experience. There is no primary aspect to that consciousness. Perceptions are blocked, and we may not respond to pain. Our best evidence indicates that tertiary aspects of consciousness are also absent. There is no volitional or executive control. There is no self-awareness, and any potential reflexive value from deep-sleep dreaming is generally lost from recall on awakening. There is autonomic "zombie-like" motor activity that can clumsily accomplish complicated but common waking tasks (e.g., opening a door or window, preparing food, walking down stairs, picking up and potentially loading a weapon for use, and running down the street). Computation is evidently not possible and combination locks can be useful for patients wanting to protect themselves from the more dangerous nocturnal behaviors possible outside their room. At this limit of consciousness there is the possibility that we will experience intense fear and extreme emotions. These emotions, which are usually negative, often occur in association with autonomic discharge (sweating, shaking, flushing, and nausea). The dreamer can become extremely angry, hitting out when woken. There are aspects of consciousness still present in deep sleep. Most commonly these are secondarily conscious images. Those underground spinning wheels are not an image derived from experience or from any memory of waking perception. They are images from archetype, intrinsic primitives that hover in deep sleep at the very edges of consciousness (see Chapter 8).

Deep sleep, deep space, and the deepest oceans are final frontiers that offer possibilities for scientific insight into organized states far different from our normal waking reality. In attempting to approach the consciousness of deep sleep, we are approaching the limits of human capabilities for analysis and understanding. We are approaching what is usually a non-conscious, basic, required state for human functioning; a state that most of us, after middle age, no longer experience: the barely comprehensible depths of deep sleep.

Notes

1. Eluard, P. (1924) Wikipedia, accessed December 22, 2012, used as an epigraph by Patrick White in *The Sacred Mandala*, as a quote by Gerald Murname in Inland, Dalkey Archive, and in a review of that book by J. M. Coetzee, The quest for the girl from Bendigo Street, *New York Review*, December 20, 2012; LIX: 60.
2. Cairns-Smith, A. G. (1996) *Evolving the Mind: On the Nature of Matter and the Origin of Consciousness*. Cambridge: Cambridge University Press, pp. 193–194.
3. Wolfson, E. (2011) *A Dream Interpreted Within a Dream: Oneiropoiesis and the Prism of Imagination*. New York: Zone Books, p. 58.
4. Koukouu, M. and Lehmann, D. (1993) A model of dreaming, in *The Functions of Dreaming*, ed. A. Moffitt, M. Kramer and R. Hoffmann. Albany, NY: SUNY Press, pp. 52–54.
5. Nielsen, T., McGregor, D., Zadra, A., Ilnicki, D. and Ouellet, L. (1993) Pain in dreams. *Sleep* 16: 490–498.
6. Pagel, J. F. and Scrima, L. (2010) Psychoanalysis and narcolepsy, in *Narcolepsy – A Clinical Guide*, ed. M. Goswami, S. R. Pandi-Perumal and M. Thorpy. New York: Springer/Humana Press, pp. 129–134.
7. States, B. (1997) *Seeing in the Dark: Reflection on Dreams and Dreaming*. New Haven, CT: Yale University Press, p. 97.
8. LaBerge, S. (1985) *Lucid Dreaming*. Los Angeles, CA: JP Tarcher. Stephen LaBerge meets all criteria for the dream researcher who has taken up fully the available role as shaman (see Chapter 2).
9. Voss, U., Holzmann, R., Tuin, I. and Hobson, J. A. (2009) Lucid dreaming: A state of consciousness with features of both waking and non-lucid dreaming. *Sleep* 32: 1191–1200. This major paper included three subjects who could volitionally signal observers.
10. LaBerge, S. (1985) *Lucid Dreaming*; LaBerge, S. (1990) *Exploring the World of Lucid Dreaming*. New York: Ballantine Books. The 1985 version is increasingly difficult to find, potentially reflecting the rumor that the author bought up all available copies, in part due to the included data indicating that lucid dreaming occurs at sleep onset and outside REMS.
11. LaBerge, S. (1985) *Lucid Dreaming*; LaBerge, S. (1990) *Exploring the World of Lucid Dreaming*.
12. Voss, U., et al. (2009) Lucid dreaming.

13. Krakow, B., Kellner, R., Pathak, D. and Lambert, L. (1995) Imagery rehearsal treatment for chronic nightmares. *Behaviour Research and Therapy* 33: 837–843.

14. Holzinger, LaBerge and Levitan, V. (2006) Psychophysiological correlates of lucid dreaming. *Dreaming* 16: 88–95; LaBerge, S., Nagel, L., Dement, W. and Zarcone, V. (1981) Lucid dreaming verified by volitional communication during REM sleep. *Perceptual and Motor Skills* 52: 727–732; Tholey, P. (1983) Relation between dream content and eye movements tested by lucid dreams; Voss, U., et al. (2009) Lucid dreaming.

15. LaBerge, S. (1985) *Lucid Dreaming*.

16. Lai, Y. and Siegel, J. (2011) Pontomedullary mediated REM-sleep atonia, in *Rapid Eye Movement Sleep – Regulation and Function*, ed. B. N. Mallick, S. R. Pandi-Perumal, R. W. McCarley and A. R. Morrison. New York: Cambridge University Press, pp. 121–129.

17. Dresler, M., Wehrle, R., Spoormaker, V. I., Koch, S. P., Holsboer, F. and Steiger, A., et al. (2012) Neural correlates of dream lucidity obtained from contrasting lucid versus non-lucid REM sleep: A combined EEG/fMRI case study. *Sleep* 35: 1017–1020.

18. LaBerge, S., et al. (1981) Lucid dreaming verified by volitional communication during REM sleep; Voss, U., et al. (2009) Lucid dreaming; Dressler M., et al. (2012) Neural correlates of dream lucidity obtained from contrasting lucid versus non-lucid REM sleep.

19. Pagel, J. F. (2011) The synchronous electrophysiology of conscious states. *Dreaming* 22: 173–191. This paper summarizes current evidence supporting the existence of a functioning electrophysiological system in the human CNS.

20. Tholey, P. (1983) Relation between dream content and eye movements tested by lucid dreams. *Perceptual and Motor Skills* 56: 875–878; Voss, U., et al. (2009) Lucid dreaming.

21. Pagel, J. F. (2013) Lucid dreaming: The conscious control of sleep offset. *Sleep Abstracts*.

22. Nietzsche, F. (1968) The birth of tragedy, in *Basic Writings of Nietzsche*, trans. and ed. W. Kaufmann. New York: New York Modern Library, Section 4, p. 44.

23. Rinpoche, T. W. (1998) *The Tibetan Yogas of Dream and Sleep*. Ithaca, NY: Snow Lion. There is a theory present among some who meditate that enlightenment is dependent on the amount of time spent meditating. Based on this perspective, sleep meditation allows the individual to increase the amount of time spent in meditation, therefore increasing the chance for enlightenment.

24. Olgivie, R., Hunt, H., Tyson, P., Lucescu, M. and Jeankin, D. (1978) Searching for lucid dreams. *Sleep Research* 7: 165; Wallace, B. A. (2006) *The Attention Revolution: Unlocking the Power of the Focused Mind*. Boston, MA: Wisdom.

25. Breton, A. (1924) *The First Manifesto of Surrealism*, in *Surrealism*, (1971) ed. P. Waldberg. New York: McGraw Hill, 66–75.; Ades, D. and Gale, M. (2007) Surrealism, in *The Oxford Companion to Western Art*, ed. H. Brigstocke. Oxford: Oxford University Press.

26. Foulkes, D. (1985) *Dreaming: A Cognitive–Psychological Analysis*. Hillsdale, NJ: Lawrence Erlbaum Associates.

27. McKellar, P. (1957) *Imagination and Thinking*. London: Cohen and West.

28. Hobson, J., Hoffman, S., Helfand, R. and Kostner, D. (1987) Dream bizarreness and the activation synthesis hypothesis. *Human Neurobiology* 6: 157–164.

29. Hunt, H. (1989) *The Multiplicity of Dreams*. New Haven, CT: Yale University Press, pp. 223–224. This is a seminal book in the field of sleep science.

30. Kaplean, P. (1980) *The Three Pillars of Zen: Teaching, Practice and Enlightenment*. Garden City, NY: Anchor Books.

31. Wolman, R. and Kozmova, M. (2006) Last night I had the strangest dream: Varieties of rational thought processes in dream report. *Consciousness and Cognition* doi: 10.1016/j.concog2006.09.009.

32. Pace-Schott, E. (2003) *Postscript: Recent findings on the neurobiology of sleep and dreaming*, in Sleep and Dreaming: Scientific Advances and Reconsiderations, ed. E. Pace-Schott, M. Solms, M. Blagrove and S. Harnard. Cambridge: Cambridge University Press, pp. 335–350; Walker, M. (2011) The role of REM sleep in emotional brain processing, in *Rapid*

Eye Movement Sleep – Regulation and Function, ed. B. N. Mallick, S. R. Pandi-Perumal, R. W. McCarley and A. R. Morrison. Cambridge: Cambridge University Press., pp. 339–349; Nielsen, T. (2003) A review of mentation in REM and NREM sleep: "Covert" REM sleep as a possible reconciliation of two opposing models, in *Sleep and Dreaming: Scientific Advances and Reconsiderations*, ed. E. Pace-Schott, M. Solms, M. Blagtove and S. Harnad. Cambridge: Cambridge University Press, pp. 59–74.

33. LaBerge, S. (1985) *Lucid Dreaming*; Voss, U., et al. (2009) Lucid dreaming.
34. Domhoff, G. W. and Schneider, A. (1999). Much ado about very little: The small effect sizes when home and laboratory collected dreams are compared. *Dreaming* 9: 139–151.
35. Dement, W. and Kleitmann, N. (1957) The relation of eye movements during sleep to dream activity: An objective method for the study of dreaming. *Journal of Experimental Psychology* 53: 339–346; Stickgold, R., Pace-Schott, E. and Hobson, J. (1994) A new paradigm for dream research: Mentation reports following spontaneous arousal from REM and NREM sleep recorded in a home setting. *Consciousness and Cognition* 3: 16–29.
36. States, B. O. (1994) Authorship in dreams and fictions. *Dreaming* 4: 237–253.
37. Pagel, J. F. and Nielsen, T. (2005) Parasomnias: Recurrent nightmares, in *International Classification of Sleep Disorders – Diagnostic and Coding Manual (ICSD)*. Westchester, IL: American Academy of Sleep Medicine.
38. Zadra, A. and Donderi, D. (2000) Nightmares and bad dreams: Their prevalence and relationship to well being. *Journal of Abnormal Psychology* 109: 273–281.
39. Pagel, J. F. (2004) Polysomnographic variables affecting reported dream and nightmare recall frequency. *Sleep – Abstracts*.
40. Pagel, J. F. and Shocknasse, S. (2007) Dreaming and insomnia: Polysomnographic correlates of reported dream recall frequency. *Dreaming* 17: 140–151.
41. Hartmann, E. (1984) *The Nightmare*. New York: Basic Books.
42. Pagel, J. F., Kwiatkowski, C. and Broyles, K. (1999) Dream use in film making. *Dreaming* 9: 247–296.
43. Pagel, J. F. and Nielsen, T. (2005) Parasomnias: Recurrent nightmares.
44. Schenck, C. and Mahowald, M. (2002) REM sleep behavior disorder: Clinical, developmental and neuroscience perspectives 16 years after its formal identification in sleep. *Sleep* 25: 120–138. RBD was first clearly identified as a diagnosis in Minneapolis by Mark Mahowald and Carlos Schenck. This accomplishment defines their careers.
45. Nielsen, T. (1999) Mentation during sleep. The NREM/REM distinction, in *Handbook of Behavioral State Control. Cellular and Molecular Mechanisms*, ed. R. Lydic and A. Baghdoyan. Boca Raton, FL: CRC Press.
46. Strauch, I. and Meyer, B. (1996) Dreams in different sleep stages, *In Search of Dreams: Results of Experimental Dream Research*. Albany, NY: SUNY Press, pp. 135–136.
47. Nielsen, T. (2003) A review of mentation in REM and NREM sleep.
48. Nielsen, T. (2003) A review of mentation in REM and NREM sleep, p. 61.
49. Domhoff, G. W. and Schneider, A. (1999). Much ado about very little.
50. Descartes, R. (1641) *Meditations on First Philosophy*, trans. D. A. Cress (1980) Indianapolis, IN: Hackett.
51. Pagel, J. F. (2003) Non-dreamers. *Sleep Medicine* 4: 235–241.
52. Solms, M. and Turnbull, O. (2002) *The Brain and the Inner World: An Introduction to the Neuroscience of Subjective Experience*. New York: Other Press, p. 184.
53. Pagel, J. F. and Shocknasse, S. (2007) Dreaming and insomnia: Polysomnographic correlates of reported dream recall frequency. *Dreaming* 17: 140–151.
54. Hartmann, E. (1994) Nightmares and other dreams, in *Principles and Practice of Sleep Medicine* (2nd ed.), ed. M. Kryger, T. Roth and W. Dement. London: W.B. Saunders, pp. 407–410.
55. Goodenough, D. (1991) Dream recall: History and current status of the field, in *The Mind in Sleep*, ed. S. J. Ellman and J. S. Antrobus. New York: John Wiley and Sons. pp. 143–171.

56. Kuiken, D. and Sikora, S. (1993) The impact of dreams on waking thoughts and feelings, in *The Functions of Dreaming*, ed. A. Moffitt, M. Kramer and R. Hoffman. Albany, NY: SUNY Press, pp. 419–476. Kuiken is another of the Canadian dream researchers who have reached far beyond those of us confined to the United States to explore the more interesting aspects of the state of dreaming.

57. Pagel, J. F. and Kwiatkowski, C. F. (2003) Creativity and dreaming: Correlation of reported dream incorporation into awake behavior with level and type of creative interest. *Creativity Research Journal* 15: 199–205.

58. Koukkou, M. and Lehmann, D. (1983) Dreaming: The functional state-shift hypothesis, a neuropsychophysiological model. *British Journal of Psychiatry* 142: 221–231. This is one of the seminal and most referenced papers in dream science.

59. Pagel, J. F. (2004) Sleep disorders in children, in *Oxford Textbook of Primary Care*, ed. R. Jones, R. Grol, N. Britten, D. Mant, L. Culpepper, C. Silagy and D. Gass, Vol. 2 – Clinical Management. Oxford: Oxford University Press, pp. 1030–1033.

60. American Academy of Sleep Medicine (2005) Arousal disorders, in *The International Classification of Sleep Disorders – Diagnostic and Coding Manual (ICSD)*. Westchester, IL: American Academy of Sleep Medicine, pp. 137–147.

61. Bonnet, M. (2000) Sleep deprivation, in *Principles and Practice of Sleep Medicine* (3rd ed.), ed. M. Kryger, T. Roth and W. Dement. Philadelphia, PA: W.B. Saunders, p. 65.

62. Pagel, J. F. (2008) Sleep and dreaming – Medication effects and side-effects, in *Sleep Disorders – Diagnosis and Therapeutics*, ed. S. R. Pandi-Perumal, J. C. Verster, J. M. Monti, M. Lader and S. Z. Langer. London: Informa Healthcare, pp. 627–642.

63. Horvitz, S., Braun, A., Carr, W., Picchioni, D., Balkin, T., Fukunaga, M. and Duyn, J. (2009) Decoupling of the brain's default mode network during deep sleep. *Proceedings of the National Academy of Sciences of the United States of America* 106: 11376–11381.

The Nightmare: Integrating Emotions into Consciousness

The mind is not a vessel to be filled, but a fire to be kindled. **(Plutarch) (1)**

The neuroanatomy of the human central nervous system (CNS) is incredibly complex. Its constituent components include over a hundred billion neurons, each with a multiplicity of cellular process interactions, most with a supporting structure of astrocytes and glial cells, and each utilizing multiple chemical neuromodulators for cellular communication (2). Yet the complex neuroanatomy and neurochemistry are only a part of the picture. Modern scanning modalities indicate that the CNS changes rapidly on a millisecond to millisecond timeline, fluxing and integrating at the speed of thought. Widely separated neural systems operate flexibly, in concert, utilizing extracellular electrical fields to coordinate their expression. At the cellular level, neuron responses extend beyond the classically perceived role as an on–off switch for conveying the transmission of signals. Each neuron is a world of its own, each containing a manipulatable metabolic system with set ionic balance and varying energy requirements, and each including a repository of biological intracellular information stored in communicating proteins and based on cellular DNA – the most complex of biological molecules. These interacting systems with their incredible multilayered complexity comprise the biological framework for consciousness. What we currently understand of these systems is quite limited, and there are likely to be other components yet to be discovered. Based on this overview, consciousness is frustratingly intangible, physiologically tied to biological systems of almost incomprehensible complexity.

The neuroconsciousness theories that attempt to explain this grand and complex process are limited by their simplicity, their reliance on one neuroanatomical or analytical perspective, and their exclusion of empirical and scientific evidence that was inconsistent with their specific approach (see the Preface). There has been a consistent tendency for neuroconsciousness theories to be concrete, poorly developed, and based on minimalist and contrived evidence. An alternative and more rational approach to studying consciousness is to use an inductive methodology that is based on available evidence rather than theoretical supposition (see Chapter 4).

In the preceding two chapters, as suggested by the principles of Descartes' scientific method (see Chapter 3), we have used the perspective of dream and sleep science to divide the problem of consciousness into component parts. Consciousness can be classified as primary, secondary, and tertiary based on evolutionary, ontological, and behavioral perspectives (see Chapter 4). In the adult human, all of these types of conscious processing are integrated into forms of waking and sleeping consciousness that have characteristic phenomenology and physiology (see Chapters 5 and 6). From this perspective, we conceptually arrive at a different view in which conscious forms come and go, change, integrate, and coalesce into an overall state that we call consciousness. During sleep, the transition between forms follows an electrophysiological pattern that can be described as a hypnogram (Figure 7.1). As noted in Chapter 6, the phenomenological forms of sleeping consciousness are state associated, rather than state restricted. During waking and at the border between waking and sleep, the integration of the forms of consciousness becomes more complex, and the transition between states less defined. Behaviors, our observable activities, are our responses to the aggregate of internal and external stimuli. These are aspects of mental processing that can be observed externally and studied.

The study of behavior led to the development of psychology as a science. The psychoanalytic movement viewed behavior as based on internal mental constructs that could not be observed or studied. In reaction, psychologists chose to study aspects of behavior that could be

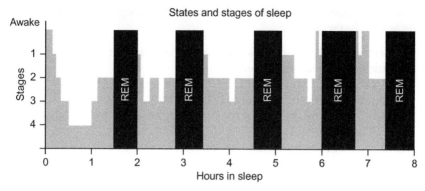

FIGURE 7.1

experimentally manipulated. They have most often studied behaviors independent of introspection and consciousness. Yet, even for behaviorists, it is clear that subjective consciousness is a component of every type of behavior. From the perspective of dream science, there is good evidence that every behavior typically includes mental processing derived from multiple forms of waking and sleeping consciousness. In this chapter, we will extend this exploration of these conscious forms of sleeping and waking into their associated emotional behaviors and feelings.

FEAR AND EMOTIONAL COMPETENCE

The emotions and feelings that affect much of our sleeping and waking activity have been the focus of scientific study since Darwin. This is an area of study that is typically expansive, involving body, brain, and mind. Emotional behavior can be morally confusing to study, "bedeviling" even for the philosophers and scientists drawn into its fascination:

> Emotion is a topic that, more than any other, has bedeviled students of mental life. The peculiar characteristics of emotional behavior and experience have been repeated stumbling blocks in the attempt to see human nature as rational, intelligent and even sublime. (*Gregory, 1987) (4)*

Emotions, agitations, or disturbances of the mind consist of two components: emotional expressions and their associated conscious sensations (feelings) (5). Various lists of the emotions include pleasure, pain, elation, euphoria, ecstasy, sadness, desire, hope, aversion, despondency, depression, fear, contentment, anxiety, surprise, anger, and hostility. These mental experiences form the "bedrock of our minds," affecting and coloring all of our actions and thought (6). Neurophysiologically, emotion has four major characteristics:

- Experiences and perceptions that acquire emotional significance can become "emotionally competent stimuli."
- Once a stimulus acquires emotional significance, a pattern of autonomic and motor responses is triggered.
- Circuits in the cerebral cortex are then triggered that process the associated feelings.
- Feedback from the peripheral, autonomic, and skeletomotor systems of emotional expression interact with conscious states of feeling in the cerebral cortex.

Like sensory perceptions and motor functioning, emotional expressions and feelings are controlled by distinct neuronal circuits. The peripheral expression of emotions is mediated in the hippocampus, an area that also plays an important role in the processing of memory. Feelings are processed in the limbic system, a group of interconnected brain sites that includes the amygdala, the area that has a coordinating role in the processing of emotional expressions and feelings. The amygdala, in turn, has strong interconnections with cortical centers in the frontal, cingulate, and parahippocampal cortices of the cerebral cortex (7). This system is a repository for neural networks that form maps of specific emotional memories. Emotion-associated feelings are our internal "perceptions" of these maps. Feelings may utilize the same CNS processing systems that are used in processing external sensations or *qualia* (sensory qualities of the environment, e.g., the color red, the smell of coffee, that are commonly experienced yet difficult to describe). We may use these neuroanatomical perceptual systems directed "internally" to perceive the "essence" of emotional memories and their associated expressions (8). This CNS emotional processing system is integrated and connected with sensory and perceptual systems; CNS memory, processing, and integrating systems; and with the extensive networks of motor, sympathetic, endocrine, and parasympathetic expression that extend throughout the body.

Dreams are characterized by the presence of emotion. In some dreams, emotion is the primary characteristic of the remembered dream. The emotions associated with dreaming are more likely to be negative than positive. Cross-culturally, and for both men and women, dream content typically includes more aggression than friendliness, and more misfortune than good fortune (9). The emotions in dreams can be profound and disturbing experiences. The intensely experienced negative emotions of anxiety, fear, terror, anger, rage, embarrassment, and disgust are typically associated with nightmares. Nightmares are commonly experienced, and

TABLE 7.1   Neuroanatomical Areas of the Central Nervous System Involved in the Processing of Fear-Associated Memories and Emotions (3)

Amygdala	Control center for processing affect load; critical area for the processing of conditioned fear, fear memory, fear detection, and autonomic activation, activated with visual perception, in PTSD, hyperactive response to trauma-related stimuli
Medial prefrontal cortex (MPFC)	Down-regulator for amygdala emotional activity, extinction memories utilized in inhibiting conditioned fears, decreased activity during traumatic imagery in PTSD
Hippocampus	Regulates the extinction and expression of conditioned fear via the amygdala and MPFC, controls the context of fear memories
Anterior cingulate cortex (ACC)	Regulates the degree of affect expression during emotional activation, involvement in the neural circuitry of pain

Note: PTSD = post-traumatic stress disorder.

even when recurrent and distressing, not necessarily the result of trauma. Intense negative emotions are also characteristically associated with other dreams and parasomnias. These include hypnogogic hallucinations at sleep onset, post-traumatic stress disorder (PTSD) nightmares and sleep paralysis that occur in sleep onset and rapid eye movement sleep (REMS), and the night terrors and confusional arousals occurring out of deep sleep.

The detection, generation, maintenance, and remembering of fear – the primary negative emotion – neuroanatomically take place in four primary areas of the CNS that operate in concert. Emotionally contextual memories (fear memories) stored in the hippocampus are relayed to the amygdala, which is interconnected with the medial prefrontal cortex and the anterior cingulated cortex. All of these areas interact with brain regions controlling sensory, autonomic, motor functioning (10). The postulated roles for each of these brain regions implicated in the processing of fear and fear-associated emotions are summarized in Table 7.1.

IMAGINING POST-TRAUMATIC STRESS DISORDER

PTSD is the only psychiatric disorder unquestionably induced by life experience. PTSD is also the major psychiatric diagnosis known to cause disruptions in dreaming. The primary symptoms of PTSD are anxiety, hyperarousal, and recurrent nightmares. These symptoms result in compromised waking function. Nightmares are the most commonly experienced of fear-associated dreams. Recurrent nightmares are the most common symptom of PTSD, present in more than eighty percent of those meeting the criteria for the diagnosis (11). These nightmares often occur at sleep onset, leading to difficulties with initiating sleep and to severe insomnia. Most of these nightmares result in arousal or awakening from sleep. While almost all nightmares occur in association with REMS, in individuals with PTSD they occur throughout sleep. Fear is a strong negative emotional stimulus that can help an individual to avoid dangerous life situations. It is difficult to postulate in the situation of PTSD that these fear-based dreams have any useful function. Freud proposed that the mental dynamics leading to what we now call PTSD develops as follows:

> We describe as "traumatic" any excitations from outside which are powerful enough to break through the protective shield. It seems to me that the concept of trauma necessarily implies a connection of this kind with a breach in an otherwise efficacious barrier against stimuli. Such an event as an external trauma is bound to provoke a disturbance on a large scale in the functioning of an organism's energy and to set in motion every defensive measure … there is no longer any possibility of preventing the mental apparatus from being flooded with large amounts of stimulus … *(Freud, 1916/1951) (12)*

Approximately twenty-five percent of individuals who have experienced severe emotional or physical trauma will develop PTSD. Nightmares that are recurrent and persistent in an individual who has experienced significant psychological or physical trauma are likely to be a symptom of PTSD. The tendency for an individual to develop PTSD is based on both premorbid personality and the severity of the experienced trauma affect (13). In war, the number of individuals who develop PTSD varies based on the nature of the conflict. Thirty percent of Vietnam War veterans were diagnosed with PTSD, while symptoms of PTSD developed in eight percent of veterans of the Gulf War. An extremely high incidence of PTSD (sixty-eight percent) in veterans of the 1973 Arab–Israeli conflict may reflect the intensity and continuous nature of that modern war experience. While this was a short-term conflict, most of the soldiers involved fought for days without rest or relief. In some civilian groups (e.g., immigrant psychiatric patients), the incidence of PTSD exceeds forty percent (14). As Freud suggested years ago, the recurrent nightmares associated with PTSD may mark a failure of those parts of the CNS involved in the emotional processing of significant life trauma (3).

PTSD nightmares can occur during REMS, but they often occur throughout sleep. PTSD is a diagnosis that clearly alters REMS, inducing increased REMS pressure (i.e., a higher tendency to go into REMS when able to sleep) (15). Imaging studies indicate that the areas processing fear and negative emotions have increased activity during REMS (Table 7.1) (16). This association between REMS and PTSD supports the contentions of the various theories that postulate a function for REMS in emotional regulation. It has been proposed that PTSD nightmares are secondary to the failure of a system of fear-memory extinction that functions during REMS dreaming. Trauma-related nightmares that persist over time and continue to generate distress could reflect a failure of this process (3). Quoting Matthew Walker, "… one of the functions of REM sleep is to tell the brain to sift through the day's events, process the negative emotions attached to them, then strip it away from memories" (17). This somewhat simplistic view of a complex postulate suggests that REMS somehow acts as a filter during sleep that is able to strain out the negative emotions of waking life experience. In the popular medical press this perspective has been extended to the suggestion that any medication that suppresses REMS is likely to be contraindicated in patients with PTSD. The medications most likely to suppress REMS include the benzodiazepines and antidepressants, yet these are currently among the few drugs that have shown positive results in the treatment of PTSD. Currently, we are faced with a suicide epidemic among soldiers suffering form PTSD who are returning home from conflicts. It seems disingenuous to suggest that drugs used to treat their anxiety and depression should be discontinued based on such an unproven theoretical construct (18).

Like REMS, dreaming is likely to have a functional role in the processing of the emotions associated with negative life experiences. Almost all dreams from all stages of sleep include emotional content. After sleep that includes dreaming, waking emotional mood is known to change. Pre-sleep mood is different from post-sleep mood, with the variability and intensity of emotion decreasing across the night (19). For many dreamers, this change in affect is in the positive direction. In individuals with PTSD, however, the change is in the opposite direction, with negative dreams producing night-time insomnia and waking distress (20). While it is possible that the processing of negative emotions during sleep is restricted to REMS, there is little other evidence supporting the theoretical postulate. Dreams can be categorized based on their amount of negative emotional content and their relationship to trauma (Table 7.2).

PTSD is difficult to treat. In many cases, it becomes a disorder with symptoms that are recurrently experienced throughout life. It is most often treated initially with acute stress debriefing that takes place on the battlefield or in the police station. This is the same "treatment" process that is used to collect legal evidence of malfeasance and is the most commonly utilized "behavioral" treatment for PTSD. For many if not most cases of PTSD this minimalistic behavioral approach is coupled with the use of medications – typically selective serotonin reuptake inhibitor (SSRI) antidepressants and antianxiety benzodiazepines (Valium-like agents) (22). All varieties and types of psychotropic medications have been used at some point in treating individuals with PTSD.

TABLE 7.2   A PTSD-Based Classification System of Dreams and Nightmares (21)

Dream Type	Definition
Dream	Reportable mental activity that occurs during sleep
Anxiety dream	Frightening dream that does not awaken the dreamer that is recalled after waking up in the morning
Nightmare	Frightening dream that awakens the sleeper
Post-traumatic dream	Dream content is associated with traumatic events by the dreamer
Post-traumatic anxiety dream	Frightening post-traumatic dream recalled only after waking in the morning
Post-traumatic nightmare	Frightening post-traumatic dream that awakens the sleeper
Replicative post-traumatic nightmare	Dream content is a replication of the original traumatic event
Symbolic post-traumatic nightmare	Dream content can be trauma related but is not a replay of the original traumatic event

Beyond the suppression of associated symptoms, it is rare for any drug to actually help in the treatment of PTSD. While there are case reports that relate positive outcomes in individual patients, it has been difficult to demonstrate that any of these medications has any consistent overall efficacy. The medications currently in common use for the treatment of PTSD (benzodiazepines and antidepressants) are older medications that have not been subjected to clinical trials in PTSD patients. Their use is based on anecdotal case reports of positive outcomes, and the tendency of patients with PTSD to develop symptoms of depression and anxiety. There are few data indicating that any medication can lead to an improvement in long-term outcome except in the treatment of those patients with significant associated depression. In these patients with PTSD and depression, thirty to fifty percent demonstrate a positive response to treatment with antidepressants (23).

The most successful therapeutic techniques for PTSD have been those designed to reduce nightmares. Some PTSD patients have abnormalities in sympathetic nervous system function, parroting patients with high blood pressure, and probably reflecting the chronic hyperarousal typical of PTSD. Such chronic sympathetic discharge can be treated with commonly used antihypertensive agents, reducing both nightmare frequency and nightmare distress. For some PTSD patients, these medications have been remarkably effective. Antihypertensive medications (primarily Prazosin) have shown excellent results in treating recurrent nightmares, reducing both nightmare frequency and insomnia (24).

PTSD-associated nightmares and insomnia respond far better to cognitive and behavioral therapies than they do to medication. PTSD has most often been treated with individual and group psychotherapy, but little evidence has been developed to demonstrate that these approaches work any better than supportive reassurance. One alternative behavioral treatment is prolonged exposure therapy that focuses on the reliving of the traumatic experience. Acutely, this approach, called "critical incidence stress debriefing" (CISD), is the recommended approach used by both military and civilian first response units. Such stress debriefing coupled with medication (antianxiety medication) has shown improved outcomes compared to medication alone (25). For some individuals, however, such acute debriefing without further support has been shown to worsen symptoms of PTSD. These symptoms can include an increase in nightmare-associated distress and insomnia (26).

Cognitive–behavioral treatments for PTSD have had the best success in treating both nightmares and insomnia. The approaches showing consistent positive results include desensitization, imagery rehearsal, and eye movement desensitization and reprocessing (EMDR). Favorable outcomes have also been seen with the newer behavioral approaches of memory restructuring and dialectical behavioral therapy (27). The cognitive component of these behavioral treatments for PTSD nightmares involves helping patients to understand that their recurrent nightmares have become habitual. Over many years, these nightmares are often utilized to maintain memories of the traumatic experience and the associated personal loss. Although distressing, they are often viewed as quite precious, and the sufferer often resists giving up the nightmare, and resents any suggestion that it is not real and significant. The best results in reducing or eliminating these nightmares have been for the behavioral therapy called "imagery rehearsal." The nightmare sufferer is asked to change the content of one of his or her nightmares, "in any way you wish," and then advised to rehearse the "new" dream while awake. This approach often leads to a decline in nightmare frequency, as well as improvements in sleep quality, anxiety, and reported levels of distress (28). This is an approach that I use clinically, as well as lucidly when personally experiencing a nightmare (see the snake nightmare reported in Chapter 6). As noted, this approach can change the import and meaning of a dream. The dreamer utilizing imagery therapy to change a nightmare often changes the dream from something realistic, frightening, and meaningful to an experience that can be something like my rewriting of my nightmare – a game with blow-up pool toys. That changed dream has its own meanings and associated memories, perhaps less distressful but sometimes just as difficult to understand and accept. The fear that I had for my wife and child in the nightmare is now sublimated within a superficially happy but deeply dissonant dream that associates with the memory of my brother's near drowning during my own childhood. Any approach to addressing nightmares in a PTSD patient works best when a therapist is involved and available for support, since addressing the emotional content can often lead to distress, and sometime to psychological decompensation.

In the situation of acutely experienced trauma, there is little evidence that any acute approach can help to keep an individual from developing PTSD (29). All of these treatment approaches attempt, using either medication or therapy, to suppress trauma-associated nightmares. Most can have at least short-term positive effects in some patients. Unfortunately, on a long-term basis, all of these approaches have demonstrated only limited efficacy. In the majority of cases, PTSD becomes a life-long anxiety disorder with fewer symptoms during easy times and exacerbations during times of stress. The chronic symptoms of PTSD are likely to persist throughout life. Among severely traumatized individuals with PTSD, such as Holocaust survivors, the recurrent nightmares of their experiences can persist into extreme old age (30). PTSD is among the most difficult of psychiatric disorders to treat effectively.

CREATIVE NIGHTMARES: HONOR, POWER, AND THE LOVE OF WOMEN

Many successfully creative individuals have frequent nightmares that they use in their work. This is but one of the findings from our work with Sundance Institute actors, directors, and screenwriters (31). Individuals in creative roles had a higher frequency of nightmares and were more likely to use their dreams and nightmares in their creative process than either the general population or the individuals on the film crew with non-creative roles. Some of these individuals had nightmares as a symptom of PTSD. For certain individuals, the incorporation of trauma-based nightmares into the creative process may be a functional therapeutic strategy for addressing the symptoms of PTSD, particularly since other approaches to treatment have proven to be of limited value.

Freud suggested that the psychological origin of creativity was in childhood trauma. This trauma was most likely to be the kind of separation and/or family relationship transition that we all experience (32). Post-Freud, a series of studies based on a review of historical records as well as retrospective psychiatric assessments of artists' work suggested that creative success was often associated with significant psychopathology. Both depression and bipolar disorder (manic–depressive illness) are quite common among successfully creative artists (33). This work has been critiqued and the findings may or may not be real. Psychiatric illness is very common: in the United States, more than forty-nine percent of the general population will have symptoms of a major psychiatric illness during their lifetimes (34). Creative artists are more likely than other individuals to leave detailed autobiographical information that, years later, is still available for retrospective interpretation. That said, among poet laureates of England, more than fifty percent had a history of suicide attempts, alcoholism, and mental breakdowns. Eighteen percent successfully completed suicide (35). Some suggest that creativity can be a form of compulsive madness that drives artists to contact their dark, demonic, and usually unconscious selves. Conversely, creativity can be an approach utilized to promote psychiatric health and even happiness (36).

Freud suggested that the creative process could be therapeutic. He proposed that beyond psychoanalysis, art offered an alternative therapeutic path for the artist:

> For there is a path that leads back from phantasy to reality – the path, that is, of art … It is well known, indeed, how artists in particular suffer from a partial inhibition of their efficiency owing to neurosis … A man who is a true artist has more at his disposal. In the first place, he understands how to work over his daydreams in such a way as to make them lose what is too personal about them and repels strangers, and to make it possible for others to share in the enjoyment of them … Furthermore, he possesses the mysterious power of shaping some particular material until it becomes a faithful image of his phantasy; and he knows, moreover, how to link so large a yield of pleasure to his representation of his unconscious phantasy, that, for the time being, at least, repressions are outweighed and lifted by it. If he is able to accomplish all of this, he makes it possible for other people to once more derive consolation and alleviation from their own sources of pleasure in their unconscious which have become inaccessible to them; he earns their gratitude and admiration and he has thus achieved through his phantasy what originally he had achieved only in his phantasy – honour, power, and the love of women. *(Freud, 1980) (37)*

Associates and followers of Freud developed this approach, and creative therapies became quite popular during the mid-twentieth century. Psychoanalysis as therapy is a one-on-one process that can be provided to only a few individuals. Art therapy and psychodrama can be conducted in group settings under the overall purview of a therapist and used to treat a far larger number of individuals. Some psychoanalytic therapists excited by this approach proposed that through the use of mass media the psychoanalytic approach could be used to treat and alter the whole of society. J. L. Moreno, in 1934, asserted, "A truly therapeutic procedure cannot have less an objective than the whole of mankind" (38). While this assertion seems a bit grandiose, today psychoanalytic perspectives have achieved their greatest distribution and resonance far away from the therapeutic couch, when used in film, print, and media to develop plot, character, and storylines (39). This concept will be addressed further in Chapter 8. However, in this twenty-first century, creative therapies such as art and psychodrama are used far less often than in the past for the treatment of psychiatric illnesses such as PTSD.

THE CREATIVE NIGHTMARE PROJECT

The following PTSD nightmare is from a series of interviews that are part of the Creative Nightmare Project, a series of interviews conducted with successful artists who have trauma-associated recurrent nightmares that they incorporate into their work.

Typically, the nightmares of PTSD are frightening and sometimes stereotypical dreams that can include the re-experience of the individual's physical and mental trauma. While PTSD nightmares can include the repetitive concrete re-experiencing of the physical and/or psychological trauma, nightmare content often varies. For at least some individuals these nightmares are a source of creative insight and inspiration. The creative use of these nightmares may offer some creative individuals the potential for reaching beyond the indigestible trauma of the experience.

The following recurrently experienced nightmare is an example of both a PTSD nightmare and the constructive paradigm of dream-based stories (from that perspective it will be discussed once again in Chapter 8). This nightmare is special, in part, in that it is reported in great detail by one of our most acclaimed screenwriters. Stewart Stern is an Academy Award-nominated and Emmy-winning screenwriter who is best known for writing *Rebel Without a Cause* (1955). Stern currently teaches a course in screenwriting at the Film School in Seattle, and works each year at the Sundance Institute Screenwriting Labs. He famously tells the young screenwriters with whom he works that "the process of screenwriting is like falling off a cliff."

In 1943, Stern was a young corporal in the infantry during a winter battle in western Europe – a battle now called the Battle of the Bulge. The experience was traumatic, miserable, and life changing. On the front lines, he sang the songs of Beatrice Lille, over and over, to mark the length of his night watch in the icy, cold trenches. In the battle, he and his best friend were separated, each presumed lost. They eventually ended up in different hospitals and were treated for months for frostbite and associated infections. There was the very real possibility that Stern might lose his legs. This interview took place on one of those rare sunny late summer days in Seattle, in a pagoda in the Zen Garden. For Stern, this dream has been a recurrent, distressing, persistent (over sixty-five-year), and life-changing experience.

I find myself in NYC, the apartment in which I lived in as a child at 430 East 86th Street, Apartment 8-B. We lived there until my parents moved while I was away at war. I'm late and trying to get across the country somehow. I'm waiting in line for a train ticket, going by train back to California where I haven't been in years. But I'm running late – fearful that I'll miss the train or fearful I don't have the money to pay for the ticket. I'm terrified. I have to get to California and get my car out of hock. I am back at the old empty apartment and I realize that my father is dead and my mother is dead and the place is empty except for the suitcases stacked in the closet, none yet packed, and I know that I should have the telephone number to Grand Central Station but I've lost it. And I don't know what to pack and there are suitcases and it's almost super-human – the effort to get even half of it in. But somehow I get to the station late and the train is about to leave and I can't find my ticket or my reservation, and there are all these suitcases and no one to help even to find the train. But somehow I get on another train, and sometimes it's a very fast plane breaking the sound barrier as we're launched upward, and the train takes me to the wrong place so I'm directed to another train that will take me back to another station where my bags are. I'm terrified. I panic. I don't know where I am. I have 50 cents. Should I use it to call the garage where I left my car for repair what seems 100 years ago? In some dreams I find the car, dusty, broken, and unusable. I'm supposed to pay and take it, but I just leave it there and decide to walk home.

But now I'm still looking for the house I deserted when I left California. I'm in country where I've never been, all up and down hills far above a place like Griffith Park. I walk down to Sunset Boulevard and I know that somewhere in the Hollywood Hills is the house I own and once lived in. Someone changes my 50 cents into dimes and I phone my exchange. They don't know me anymore. I'm off their list. I pray for them to call a friend but they can't connect me and my folks don't live any more at the old number I had. Finally I climb the hills again and find my house and it's in terrible disrepair. At one time it had a view out over the city, back before there were tall buildings, and you could see all the way to the train station and Olvera Street. Then there's my house, but across the road from where I'd left it. It shocks me to see it, and I'm so ashamed at having let it go to ruins. There's broken plaster and wallpaper hanging loose from the walls. And a room I've never seen before that's such a mess and maybe used to be my den. I have no one's phone number or address here either and no way to contact anyone. I go into my house. It's full of many new rooms. I look downstairs across the banister and below me there's a movie being shot in my living room. In order to get the best shots, they've taken out the whole front of the living room that overlooks the city. The crew is all young women except for the guys pushing equipment. There is a row of hospital cots set up below me and there are actors in these cots pretending to be wounded, part of a dolly shot in which the camera moves very slowly across the front of the room above the heads of the wounded guys. And I think I see this one guy, my friend Jim from the war.

I look down at the bunks and the men in the cots are no longer actors. They're the men that were with me in the hospitals after the battle, Jim included, I think. Well, they were actors in this dolly shot in my dream playing roles as soldiers in my living room. Then there's this one guy who makes eye contact with me. He looks at me and I look at him and I have this sense of homecoming, as if I'm dreaming of Jim, but I realize that he's Heath Ledger, just an actor who I never met, but our eyes are locked together, and he is going to die the next week. And I look around the room, thinking that these women should be held accountable! I gave no one permission to shoot a movie here! Somebody owes me money! My house has been destroyed. So I once again have come home without really coming home and this dream is the model for repeated dreams of missing trains and ocean liners like the one that took me to war, and of not looking for my mother and knowing that she is somewhere in the city and I haven't tried to call. And there's a dream of a slide off an endlessly high and scary chute with no banister and I see her playing cards at the bottom. I'm terrified and Joanne Woodward is at my side urging me to slide. My mother and father don't see me land. They don't seem dead but I know they are. And there's a theatre near their table where Beatrice Lillie is playing in a comeback performance. I know that she's dead too, but she's like an automaton approximating her voice. She sees me with no recognition. And the songs are so sad.

Stern came back from the war to become one of our great storytellers, but from this transcription, it may be difficult to convey the depth of the emotions present in his telling of this dream. It is a true nightmare. There are despair, loss, failed responsibility, death, destruction, lack of control, a multiplicity of meaningful associations, and no possibility of resolution or a way out. This dream extends into his waking life, becoming especially recurrent during episodes of life stress and periodic bouts of writer's block. He has used portions of this dream in screenplays. This recurrent terrible dream has informed his extraordinary success.

Encouraging a creative process such as screenwriting could be used as part of an approach to the treatment of PTSD in which the individual uses his or her creative process to create a product that is based in part on the experience of trauma. A work of art, existing independently of the individual, allows for at least some degree of distance and dislocation. The participants in our study who had an already developed creative process at the time of the trauma often responded to the experience of trauma by changing not only their art, but also almost everything about their lives. After he was released, the captured, single, war photojournalist in our study married, had children, and became a teacher and photographer of monks. The Vietnamese boatperson who watched her father swept overboard at sea now teaches returning and similarly damaged soldiers to paint and helps them to create images based on their trauma. The young artist, a victim of serial and socially supported rape, works as a curator, organizing the art of others in remarkable patterns of presentation that often shock the viewer. This approach has been more beneficial for some than for others. For some of our participants it was difficult to demonstrate that they have in any way been able to integrate their trauma-associated nightmares into their work. The videographer who had broken his neck in a helicopter crash during filming remains unable to achieve the quality or satisfaction with his work that he had been able to attain before the trauma.

EMOTIONAL CONSCIOUSNESS

Dreams and nightmares are full of emotion. That emotion can be profound and distressing, the primary characteristic of parasomnias such as nightmares, night terrors, sleep paralysis, and hypnogogic hallucinations. Despite their involvement in emotional processing, sleep and dreaming are not usually considered in general texts or in popular scientific literature on emotion (40). When sleep and dreaming are addressed, emotions are most often discussed as if they are only associated with REMS. Antonio Damasio, who writes prodigiously on emotion, takes this further, basing theoretical constructs on his perceived understanding that, "Deep sleep is not accompanied by emotional expressions ... ." Based on this perspective, he argues that impairment in both consciousness and emotion during deep sleep reflects a disruption in our core consciousness (41). He continues to believe that REMS is the dreaming state, oblivious to the intense emotions characterizing the arousal disorders of deep sleep (42). Once again, another complicated process, the emotional processing occurring during sleep, has been reduced to a simple construct: emotion equals dreams equals REMS. Yet the emotional processing systems of our waking hours are clearly active throughout much of sleep. Emotions and their associated feelings affect all of our behaviors, our waking function, and our sleep. Individuals who have difficulty processing emotions during wakefulness have those same difficulties during sleep.

While the neuroanatomical systems of emotional processing are more active during REMS, there is no clear approach that can be used to determine whether this activity is associated with dreaming. The study of dreaming – even of negative dreams – requires a waking report that dreaming has taken place. It is clear that emotional process systems are active during emotional dreaming and dream-associated parasomnias. Intense emotions are present in the parasomnias from all of the sleep stages. For some individuals, even low-intensity stage 2 dreams can include intense emotions. Panic attacks, most commonly experienced during sleep in stage 2, are often experienced as brief moments of terror (43). It is very unusual for a dream not to include emotion. Emotion is clearly one of the primary cognitive components of all forms of dreaming.

As Freud suggested, symptoms characteristic of PTSD most often occur after the experience of a trauma of such indigestible intensity as to break down our normally effective "protective shields." The association of nightmares with PTSD suggests that the emotional processing functioning during normal dreaming is overwhelmed by such trauma. The result is recurrent, distressing nightmares. Since emotional dreams occur in all stages of sleep, it is doubtful that emotional neural processing is limited to REMS. PTSD nightmares often occur outside REMS, supporting the contention that emotional processing during sleep is not restricted to REMS. It is likely that PTSD nightmares are secondary to an individual's inability to process severe and overwhelming experiences of trauma. That emotional processing does not appear to be limited to a particular stage of sleep.

A primary approach to the treatment of PTSD is to attempt to reduce and alleviate the distress associated with the nightmares. On an acute and short-term basis, this has proven an effective therapy. Unfortunately, for many, the symptoms of PTSD become chronic waxing and waning in intensity over the persisting course of their life. Based on the interviews conducted as part of the Creative Nightmare Project, it is clear that some successful creative individuals have discovered credible paths that integrate their trauma-associated nightmares into their life's work. I doubt that any of those interviewed would say that their experience of trauma was a positive thing. I doubt that any would say that their life paths have been easy. And I doubt that any would say that they enjoy their nightmares. But some of those who have integrated their nightmares into their work have created remarkable works of art: art in the image of their "phantasies"; images that can, for a time at least, lift the weight of the experience of trauma, making it possible for others to derive "consolation and alleviation" of their own traumas in their experience of that art.

Notes

1. Plutarch, M. (48.) On listening in lectures, part of Moralia, included in *Plutarch's Lives*, Vol 1. (2004) A. Stewart, G. Long (trans.) Online reference accessed 11/24/13.
2. Swartz, J. (2000) Neurotransmitters, in *Principles of Neural Science* (4th ed.), ed. E. Kandel, J. Swartz and T. Jessell. New York: McGraw-Hill, pp. 280–297.
3. Levin, R. and Nielsen, T. (2007) Disturbed dreaming, posttraumatic stress disorder, and affect distress.
4. Gregory R. (ed.) (1987) *The Oxford Companion to the Mind*. Oxford: Oxford University Press, p. 219.
5. *The Compact Oxford English Dictionary* (1971). Oxford: Oxford University Press, p. 853.
6. Damasio, A. (2003) *Looking for Spinoza – Joy, Sorrow and the Feeling Brain*. Orlando, FL: Harcourt, p. 3.
7. Iverson, S., Kupfermann, I. and Kandel, E. (2000) Emotional states and feelings, in *Principles of Neural Science* (4th ed.), ed. E. Kandel, J. Swartz and T. Jessell. New York: McGraw-Hill, p. 982.
8. Damasio, A. (2003) *Looking for Spinoza*, pp. 86–87.
9. Domhoff, G. W. (2003) *The Scientific Study of Dreams: Neural Networks, Cognitive Development, and Content Analysis*. Washington, DC: American Psychological Association, p. 26.
10. Maren, S. and Quirk, G. (2004) Neuronal signaling of fear memory. *Nature Reviews: Neuroscience* 5: 844–852.
11. Lavie, P. (2001) Sleep disturbances in the wake of traumatic events. *New England Journal of Medicine* 345: 1825–1832.
12. Freud, S. (1916/1951) *Beyond the Pleasure Principle*, Vol. 18, The Standard Edition, trans. and ed. J. Strachey. London: Hogarth.
13. Dunn, K. K. and Barrett, D. (1988) Characteristics of nightmare subjects and their nightmares. *Psychiatric Journal of the University of Ottawa* 13: 91–93; Levine, R. and Fireman, G. (2002) The relationship of fantasy proneness, psychological absorption, and imaginative involvement to nightmare prevalence and nightmare distress. *Imagination, Cognition and Personality* 21: 111–129.

14. Ekblad, S. and Roth, G. (1997) Diagnosing posttraumatic stress disorder in multicultural patients in a Stockholm psychiatric clinic. *Journal of Nervous and Mental Disease* 185: 102–107; Fawzi, M. C. (1997) The validity of posttraumatic stress disorder among Vietnamese refugees. *Journal of Trauma and Stress* 10: 101–108; Yehuda, R. and McFarlane, A. C. (1995) Conflict between current knowledge about posttraumatic stress disorder and its original conceptual basis. *American Journal of Psychiatry* 152: 1705–1713.

15. Singareddy, R. K. and Balon, R. (2002) Sleep in posttraumatic stress disorder. *Annals of Clinical Psychiatry* 14: 183–190.

16. Nofzinger, E. (2004) What can neuroimaging findings tell us about sleep disorders. *Sleep Medicine* 5(Suppl. 1): S16–S22.

17. Walker, M. P. (2009) The role of sleep in cognition and emotion. *Annals of the New York Academy of Science* 1156: 168–197; Staples, T. (2009) Wish fulfillment? No. But dreams do have meaning. *Time* June 15, http://www.time.com/time/health/article/0,8599,190451,00.html (last accessed November 8, 2009). It is my sincere hope that PTSD patients have not stopped what were otherwise effective medications based on this theory. For better or worse, this paper has affected the tendency of therapists treating PTSD to use REMS suppressant medications in their treatment.

18. Robertson, M., Humphreys, L. and Ray, R. (2004) Psychological treatments for posttraumatic stress disorder: Recommendations for the clinician based on a review of the literature. *Journal of Psychiatric Practice* 10: 106–118; Putnam, F. W. and Hulsmann, J. E. (2002) Pharmacotherapy for survivors of childhood trauma. *Seminars in Clinical Neuropsychiatry* 7: 129–136.

19. Kramer, M. (2007) The functions of dream, in *The Dream Experience: A Systemic Exploration*. New York: Routledge, p. 183.

20. Zadra, A. and Donderi, D. (2000) Nightmares and bad dreams: Their prevalence and relationship to well-being. *Journal of Abnormal Psychology* 109: 273–281; Pagel, J. F. and Nielsen, T. (2005) Parasomnias: Recurrent nightmares *The International Classification of Sleep Disorders – Diagnostic and Coding Manual (ICSD)*. Westchester, IL: American Academy of Sleep Medicine.

21. Schreuder, B., Igreja, V., van Dijk, J. and Kleijn, W. (2001) Intrusive re-experiencing of chronic strife or war. *Advances in Psychiatric Treatment* 7: 102–108. This scale is not in common use, but provides, in an excellent descriptive sense, the range of experience associated with PTSD dreaming.

22. Pagel, J. F. (2010) Drugs, dreams and nightmares, in *Dreaming and Nightmares*, ed. J. F. Pagel, *Sleep Medicine Clinics* 5(2). Philadelphia, PA: Saunders/Elsevier; Pagel, J. F. (2010) Sleep-associated cognitive side effects of psychoactive medications, in *Sleep and Mental Illness*, ed. S. R. Pandi-Perumal and M. Kramer. Cambridge: Cambridge University Press, pp. 309–325.

23. Bisson, J. and Andrew, M. (2005) Psychological treatment of post-traumatic stress disorder (PTSD). *Cochrane Database Systematic Reviews* (2) CD003388; Putnam, F. W. and Hulsmann, J. E. (2002) Pharmacotherapy for survivors of childhood trauma.

24. Southwick, S. M., Krystal, J. H., Morgan, C. A., Johnson, D., Nagy, L. M. and Nicolaou, A., et al. (1993) Abnormal noradrenergic function in posttraumatic stress disorder. *Archives of General Psychiatry* 50: 266–274; Raskind, M. A., Peskind, E. R., Kanter, E. D., Petrie, E. C., Radant, A. and Thompson, C. E., et al. Reduction of nightmares and other PTSD symptoms in combat veterans by prazosin: A placebo-controlled study. *American Journal of Psychiatry* 160: 371–373.

25. Everly, G. and Boyle, S. (1999) Critical incident stress debriefing (CISD): A meta-analysis. *International Journal of Emergency Mental Health* 1: 165–168.

26. Krakow, B., Melendrez, D., Pedersen, B., Johnston, L., Hollifield, M., Germain, A., et al. (2001) Complex insomnia: Insomnia and sleep-disordered breathing in a consecutive series of crime victims with nightmares and PTSD. *Biological Psychiatry* 49: 948–953.

It is also recommended that patients with nightmares should avoid rehearsing these unpleasant dreams on waking and attempt to move to more pleasant thoughts.

27. Shapiro, F. and Forrest, M. (1997)   *EMDR (Eye Movement Desensitization & Reprocessing): The Breakthrough "Eye Movement" Therapy for Overcoming Anxiety, Stress, and Trauma.* New York: Basic Books.

28. Germain, A. and Nielsen, T. (2003) Impact of imagery rehearsal treatment on distressing dreams, psychological distress, and sleep parameters in nightmare patients. *Behavioral Sleep Medicine* 1: 140–154.

29. Mellman, T. A., Bustamante, V., David, D. and Fins, A. I. (2002) Hypnotic medication in the aftermath of trauma. *Journal of Clinical Psychiatry* 63: 1183–1184.

30. Sadavoy, J. (1997) Survivors. A review of the late-life effects of prior psychological trauma. *American Journal of Geriatric Psychiatry* 5: 287–301.

31. Pagel, J. F. and Kwiatkowski, C. F. (2003) Creativity and dreaming: Correlation of reported dream incorporation into awake behavior with level and type of creative interest. *Creativity Research Journal* 15: 199–205; Pagel, J. F., Kwiatkowski, C. and Broyles, K. (1999) Dream use in film making. *Dreaming* 9: 247–296.

32. Freud, S. (1908) Creative writers and daydreaming, in *Introductory Lectures on Psychoanalysis, The Standard Edition (Lecture XXIII – The Paths to the Formation of Symptoms)*, ed. and trans. J. Strachey (1966). New York: W. W. Norton & Co., p. 375.

33. Jamison, K. (1993)   *Touched with Fire: Manic–Depressive Illness and the Artistic Temperament.* New York: Simon & Schuster.

34. Kessler, R. C., McGonagle, K. C. and Zhao, S. (1994) Epidemiology of psychiatric disorders. *Archives of General Psychiatry* 51: 8–19.

35. Martindale, C. (1990)   *The Clockwork Muse: The Predictability of Artistic Change.* New York: Basic Books.

36. Kavaler-Alder, S. (2000)   *The Compulsion to Create – Women Writers and Their Demon Lovers.* New York: Other Press.

37. Freud, S. (1908) Creative writers and daydreaming, pp. 375–377.

38. Moreno, J. (1934)   Who Shall Survive? (153, p. 3), quote referenced in *Psychodrama: Theory and Therapy* ed. I. Greenburg. New York: Behavioral Publications., p. 480. (1974).

39. Pagel, J. F. (2008)   *The Limits of Dream – A Scientific Exploration of the Mind/Brain Interface.* Oxford: Academic Press/Elsevier.

40. Kalat, J. and Shiota, M. (2012)   *Emotion* (2nd ed.). Belmont, CA: Wadsworth Cengage Learning.

41. Damasio, A. (1999)   *The Feeling of What Happens: Body and Emotion in the Making of Consciousness.* San Diego, CA: Harcourt, pp. 100–106. Damasio suggests that when "emotions" are evoked by imagery rather than external stimuli, the peripheral responses and expression of that emotion (mood state) are much less marked and distinctive. Among behaviorists, dreams are generally viewed as noise-driven REMS-based processes that are based on ongoing day-concerns such as anxieties and desires.

42. Damasio, A. (1999) *The Feeling of What Happens*, p. 117.

43. Kandel, E. (2000)   Disorders of mood, depression, mania and anxiety disorders, in *Principles of Neural Science* (4th ed.), ed. E. Kandel, J. Swartz and T. Jessell. New York: McGraw-Hill, pp. 1221–1222.

Organizing Consciousness into Stories

All that we see or seem
Is but a dream within a dream **(E. A. Poe 1849) (1)**

Dreams are inherently self-organized. A dreamer requires no training to learn to present a dream as a narrative. As dreams progress, they develop a framework that includes the preconditions for a narrative or story structure. There is a logical sequence of associations, a situational dynamic in which each dream incidence occurs in response to incidents that are already there in the dreamer's memory of waking experience. Dreams have typical storylines with narratives reflecting universal concerns, much as in literature. The almost universally experienced dreams are those of flying, chase and attack, drowning, poor test performance, nakedness, and the experience of being trapped (2). Dreams foreshorten and expand these narratives, a process that is also typical of waking stories. In dreaming, these stories are in constant motion, unstable, subject to instant revision and expansion. Plot in literary fiction is also a continually evolving pattern of imagery and events (3). Yet dreams may not even have plots. Dreams, because they are intrapersonal experiences, can disregard the requirements of communicability and performance that are necessary for most stories. It is as if dreams are trying to become genuine stories but are typically falling a bit short (4).

Dreams have literary form. Dreaming and literature utilize the same patterns of story organization. These structural forms used to organize narrative are probably the essential combinational strategies that we use to organize thought and experience. The mind organizes the thoughts and memories of the dream into narrative structures using some of the same approaches that we use in waking to describe perceptual experience. These literary forms are our ways of bracketing the world. Dreams and fiction connect previously unconnected matrices of experience (5). Both recreate human experience as a narrative form that deals in hypothetical or imaginary events. Works of fiction can be viewed as waking dreams designed for other people. Dreaming, whether shared or not, connects an individual's internal particularized knowledge with general shared knowledge of the external world. As in literature, dream stories most often reflect universal concerns. For the writer focused on telling a story, dreams can supply an invaluable and nightly renewable source of input into that process.

There are dreams reportedly experienced as fully realized literature. The most famous of these instantaneous writings is Coleridge's poem *Kubla Khan*. On awakening from a three-hour nap Coleridge wrote down the first fifty-four lines. A knock at his door from a salesman interrupted, and the remaining lines of his poem were lost. However, Coleridge's claim is suspect. He had an addictive personality, and this poem was written during a period of his heaviest opium use, and likely to have been composed in a drug-induced haze rather than in a dream. Recently, earlier drafts of this poem have been discovered, making his claim even more suspect (6).

Dreams can lead to magnificent works of literature. Some of the finest works by our most honored poets, authors, and screenwriters were inspired by dreams. There is an extensive list of authors who have acknowledged the incorporation of their dreams into their work, including *La Henriade* by Voltaire, *Frankenstein* by Mary Shelley, *The Strange Case of Dr. Jekyll and Mr. Hyde* by Robert Louis Stevenson, *All the King's Men* by Robert Penn Warren, *The Dharma Bums* by Jack Kerouac, *The Milagro Beanfield War* by John Nichols, Franz Kafka's *In the Penal Colony*, Graham Greene's *Honorary Consul*, Stephen King's *It*, Sherman Alexie's *Indian*

Killer, and John Sayles' *Brother from Another Planet* (7). As an example, here is Mary Shelley's nightmare of Frankenstein:

> I saw the pale student of unhallowed arts kneeling beside the thing he had put together – I saw the hideous phantasm of a man stretched out, and then on the working of some powerful engine, show signs of life, and stir with an uneasy, half-vital motion. Frightful it must be; for supremely frightful would be the effect of any human endeavor to mock the stupendous mechanism of the Creator of the world ... He would hope that left to itself, the slight spark of life that he had communicated would fade; that this thing which had received such imperfect animation would subside into dead matter, and he might sleep in the belief that the silence of the grave would quench forever the hideous corpse which he had looked upon as the cradle of life. He sleeps; but he is awakened; he opens his eyes; behold, the horrid thing stands at his bed side, opening his curtain and looking on him with yellow, watery, but speculative eyes ... *(Shelley, 1818)* (8)

For those who write fiction, such dreams can be an almost indispensible source of inspiration.

In a dream we come as close as consciously possible to observing the "thinking of the body." In reporting dreams, we use language and narrative to structure and organize our internal (non-perceptually based) thoughts and imagery so that they can be shared with others. This is not something that an individual has to learn or be taught. Once children develop the capacity for language and communication, they begin to report dreams. Often these are the classic dreams of monsters under their beds or in their closets. Far more frequently than adults, they experience the terrors and confusional arousals that, along with sleepwalking, originate in deep sleep. Even these extreme dreams from within the farthest depths of consciousness are organized into a descriptive format that becomes a story. We use words which, combined with grammar, achieve the organized narratives that are presented as dreams. These dream narratives are stories that can be shared. Dreams are an exercise in pure storytelling that are nothing more (or less) than the organization of internal experience into set patterns that describe the mechanisms in which we order our thinking system (9). When we examine the signs, inner grammar, and narrative structure of a dream, we are observing the inner structure of the mind, and how these mechanisms operate in interpreting experience (10).

SIGNS AND ABSOLUTE SYMBOLS

The organization of thought and language is among the oldest topics to be subjected to philosophical discourse. Aristotle pointed out that words are the proof that one has something in one's mind to express (11). If words are to be shared, an understanding of both language and meaning is required so that rational communication can take place. As the scholar Thomas McGrath writes, "In the beginning was the world, and the word when it did come, finally allowed humans to share with one another the overheard secrets of the world" (12). For communication to occur, words require definitions. Aristotle viewed definitions as "the essence or essential nature" of words (13). Even for a dream, it is necessary to define the words if they are to be used in communicating the story of that dream with another.

The study of words and their meanings is referred to as *semiotics* or the study of "signs." Historically, semiotics dates back at least to the ancient Greeks. Today, it remains a powerful aspect of philosophy, contributing to the analytic approaches of Eco, Chomsky, and Derrida. Throughout the centuries, the study and analysis of signs have become exceedingly complex; a labyrinth of metaphor, symbol, and codes that can lead to an infinite set of interpretations (14).

A variety of philosophical definitions exist for "sign," most developed around the Pierce concept of "something which stands to somebody for something in some respect or capacity" (15). While diffuse, this definition is more comprehensible than the classical philosophical definition developed by Hobbes: "a sign is the evident antecedent of a consequent or the consequent of an antecedent when similar consequences have previously been observed" (16). In semiotics, a sign includes characteristics of the signifier and the signified (the expression and the content, or the name and the thing) (17). For philosophers, the study of linguistic representation using the concept of "sign" is a common approach used to address the even more difficult concept of "meaning."

Freud adapted this philosophical approach to the study of "signs" into psychoanalytic dream interpretation. For Freud, the presented (manifest) dream is a marker, a point that includes the presented words (signs) of the dream that connect to a set of associated memories, aggregations, and disintegrations that are open to an infinite set of further combinations. In Freudian dream analysis, the manifest signs presented by the dreamer are a part of a continuum of interpretation. Pushed to the extreme, as in psychoanalytic therapy, the original dream content can disappear, lost in a complex integration of infinite associations, and sometimes global segmentation (18). Psychoanalytically, the presented dream serves as an extendible metaphor that can be used by the therapist to construct representations of an individual's psyche. Unfortunately, such interpretations are often based on a theory-constitutive metaphor developed at a logical remove from the subject. Psychoanalytic dream interpretation is more likely to reflect the theoretical construct of the therapist than the patient's presented and life-associated dream content.

Most dream interpretation focuses on the shared content common to most dreams, content that can be metaphorically described.

These shared concepts are generally referred to as "symbols." In classical tradition, the first symbols were broken coins, two halves of the same thing, one standing for the other. Freud classified dream symbols as either idiosyncratic (based on the patient's life and associations) or unconscious (based on the shared ideation of myth, folklore, legends, linguistic idioms, proverbial wisdom, and current jokes) (19). Freud developed a code for interpreting the meaning of dream symbols that allowed for selected features and expressions to be used by trained therapists to understand the structure of the psyche, as well as to diagnose psychiatric illness.

Carl Jung, while appreciating many of Freud's insights, had difficulty with his mentor's objective codification system and his clinical, externally applied approach to dream interpretation:

> Now if one begins, as the Freudian school does, by taking the manifest content of the dream as "unreal" or "symbolical," and explains that though the dream speaks of a church-spire it really means a phallus, then it is only a step to saying that the dream often speaks of sexuality but does not always mean it, and equally that the dream often speaks of the father but really means the dreamer himself. ... The whole dream-work is essentially subjective, and a dream is a theatre in which the dreamer is himself the scene, the player, the prompter, the producer, the author, the public and the critic. *(Jung, 1948) (20)*

Jung extended the Freudian concept of symbols based on myth and folklore into anthropology, art, and comparative religion, "Hence we may expect to find in dreams everything that has ever been of significance in the life of humanity" (21). He theorized that on a deeper, innate level of the human psyche there were collective symbols (archetypes) that are the same everywhere and in all individuals. Jung's typical archetypes were lunar, solar, vegetal, and meteorological representations present culturally in art and myth, and repeated intrapersonally in dreams and visions. These archetypes can be mystical, indefinable, and infinitely interpretable. The circle, culturally fully realized as the Hindu/Buddhist Mandela, is a classic example of the Jungian archetype. It can be interpreted as the seat and center of the soul, the concealed "pole" around which everything ultimately revolves.

Potentially, therapist and subject could work together to arrive at the absolute metaphors that constitute the dream – characteristics that are not further reducible with conceptual thought or language. Such irreducible forms can be difficult to discover. Dream content constantly morphs from one associative memory to another during dreaming as well as during the dream report to the therapist. The circle, the soul, is as close as an archetype can come to being absolute. In his clinical work, Jung dissected even this apparently absolute metaphor into classifications of both personality and expressed behaviors. In his own journals, rather than using and explaining, he delved deeper into the unexplainable – absolute metaphors, both shared and private, that suggested everything, and perhaps meant nothing (22). Beyond content, absolute metaphors can include the formal narrative structure of the dream and grammar, as well as non-contextual imagery. An example is the "dream within a dream," a concept that can be alluded to but not metaphorically described in any other fashion (23).

SIGNING "DREAM"

Consider the meaning and definition of the sign "dream." The primary dictionary definition is derived from Aristotle – mental activity occurring during sleep. But that definition is only one of a bewildering diversity of definitions for dream that are dependent on an individual's background and approach to knowledge. This confusion of definitions pre-dates science and extends beyond the difficulties that the scientific community has with such a simplified incongruity as REMS equals dreaming (Box 8.1). Each of us who dream has a concept of what a dream is. Yet the awareness, experience, and concept of what is and is not a dream range into topics as diverse as REMS, personal motivation, neuroanatomy, film, hallucinations, literature, neuropharmacological effects of medication, real estate, the unconscious mind, and anthropology. Google "dream" or look it up at your local library. Today, the most common usage is loosely based on the Freudian concept of "wish fulfillment": dream home, dream vacation, dream marriage. In searching for its essence, the concept of "dream" becomes increasingly diffuse. Researchers have often avoided the issue by refusing to supply a definition (of 100 articles published between 1991 and 2001 in *Dreaming*, the journal for the scientific study of dreaming, only fourteen percent included a definition for "dream") (25). Yet, what a dream is to a sleep medicine physician (mental activity occurring during sleep) is very different from what a dream is for a student of psychoanalysis (bizarre or hallucinatory mutation associated with consciousness). These scientific definitions are incompatible. For the sleep physician, dreams occur in sleep; for the psychoanalyst they also occur when we are awake. For the psychoanalyst, a dream has a particular type of content; that constraint is unapparent to the sleep physician.

So what is a dream? At the turn of this century a group of dream "experts" from a wide spectrum of disciplines was recruited from the membership of the International Association for the Study of Dreaming (iASD) to address the topic. They agreed that a single definition for dreaming is impossible. This consensus group went further to develop a classification system for organizing operational definitions of "dream"

BOX 8.1

DICTIONARY DEFINITION OF DREAMING, CIRCA 1947 (24)

Dream – n. 1) a succession of images or ideas present in the mind during sleep. 2) the sleeping state in which this occurs. 3) an object seen in a dream. 4) an involuntary image occurring to one awake. 5) a vision voluntarily indulged in while awake. 6) a wild or vain fancy. 7) something of an unreal beauty or charm. – *v.* 8) to have a dream or dreams. 9) to indulge in daydreams or reveries. 10) to think or conceive of something in a very remote way. 11) to see or imagine in sleep or in a vision. 12) to imagine as if in a dream; fancy; suppose. 13) to pass or spend (time etc.) in dreaming (often followed by away).

(Box 8.2). Three converging characteristics common to dreaming can be used to construct the outline of a definition: dreams occur in the spectrum of conscious states that we describe as sleep and waking; dreams are recalled and reported in waking; and dreams include some type of content. In other words, dreams are actually a labyrinth of meanings, an associative nexus of memories, a series of oneiric symbols, a Porphyrian tree of grammatical structure, meditative visions, imagery, REMS, emotions, and day-residues, as well as an awareness of consciousness, messages from God, meaningless babble, neuroconsciousness, the borders of mind, and absolute metaphor. Faced with this contradictory and scattered spectrum of definition, the consensus group made the unpopular suggestion that any author writing about the topic should state the operational definition of "dream" being used in the text so that work could be compared with other work that addressed the same or similar topics. This seemed a rational suggestion, but the results have not all been positive. The dream science field has dispersed into specific areas of focus. Today, most general publications on "dreaming" are written by authors from outside the field who continue to adopt an intuitive approach to dreaming. For most of those who are not dream scientists, "dreaming" as personally understood is the same for everyone.

Currently in the field (since work that does not define the topic of study is often excluded), the published books on dream science are much shorter than the predefinition tomes. In this book, dreaming is defined as sleep mentation that is remembered or reported on awakening. Dreamlike phenomena are those occurring in sleep or when awake that share characteristics with dreams, particularly those reported from states that are on the border of sleep and waking (see Chapters 5 and 6).

BOX 8.2

DEFINITIONS OF DREAMING: A CLASSIFICATION SYSTEM PARADIGM (26)

A definition of a dream has three characteristic continua: (a) wake/sleep, (b) recall, and (c) content:

a. Sleep – Sleep onset – Dream-like states – Routine waking – Alert wakefulness
b. No recall – Recall – Content – Associative content – Written report – Behavioral effect
c. Awareness of dreaming – Day reflective – Imagery – Narrative – Illogical thought – Bizarre hallucinatory

---

The study of dream signs and content has become increasingly constrained by this confusion of definition and meaning. In research, there are transference effects that affect the interaction between the dreamer and the recorder. The presented dream report varies based on gender, expectations, location of collection, and any bias or perceived preunderstanding of dreams. It is methodologically difficult to control for these effects. Tightly controlled studies using validated questionnaires that are coded and then computer analyzed demonstrate that much of what we thought we knew about differences and characteristics of dream content may be incorrect (27). The same issue that has recurrently confronted psychoanalysis, semiotics, and philosophy is now confronting dream scientists. In the introductory presentations for these other areas of study, it has become traditional for instructors to make the announcement that their respective philosophy has died, that Freud is dead, or that the study of signs has moved beyond comprehension. Faced with such a conundrum, students and faculty must concretely face their fields' as well as their own inherent limitations, and then move on (28). This philosophical juncture is likely to confront most developed and cognitively complex areas of study. After at least 4000 years of human consideration, the study of dream content, the indefinable and all-inclusive signs reported as dreams, finds itself at such an impasse. The ancient process of dream interpretation is based on theoretical constructs and codes that can block both the therapist's understanding of the individual and the individual's understanding of a dream. Many dream signs are individual and tied to associated personal experience and memories. Your dream horse means something different to you from what my dream horse means to me. Bizarreness in dream behavior varies based on the questionnaire used and the bias of the researcher (see Chapter 6). Because of methodologically uncontrollable transference, "experts" on dreaming are forced to confront the words and content of dreams with computerized input and analysis in order to obtain consistent and repeatable results.

But there are aspects of dreams that lie beyond this labyrinth of content. In dreams, consciousness is organized into its formal structure. Imagery can be free of context. Rhetoric exists independent of any therapist's replacement, interpretation, or analysis of its associations. Free of signs and interpretation, surprisingly, this dimension of dreaming is not a dry and arid place. There, we find the extraordinary structures inherent in dream, absolute forms that contribute to the power of stories and images, and project out from dreaming into the symbols of shared myth.

DREAM LANGUAGE

With language we organize internal thoughts so that they can be shared. Sifting through thousands of alternative patterns and languages available for communication, linguists have worked to identify the basic underlying structures of languages as diverse as Chinese, English, Icelandic, Swahili, Basque, and Navajo. Noam Chomsky refers to the shared "internal" constructs attributable to all languages as an "I-language." This basic language faculty is a species-specific capacity. Every child is able to learn the languages to which he or she is exposed. A partial set of "principles and parameters" characterizing this I-language has been identified based on experimental studies of language perception and production, the analysis of language acquisition and change, and the study of language alterations induced by pathology. I-language properties include the ability to conceive of the property of discrete infinity, the concept of three-word sentences, and a capacity for adapting processes of generative grammar to profoundly different language formats so that expressions can have precise semantic representations (29).

We interact with the external perceptually based world by using a linguistic system based on articulatory interaction (the ability to speak and be heard). This part of the linguistic system incorporates phonetic (sound-based) representations of words. The dream excludes, at least in part, these articulatory-perceptual components of language. The linguistic pattern used to organize non-perceptual cognitive constructs such as dreams is less likely to use a "surface" syntax based on word sounds derived from the external world experience. Dreams are primarily comprised of a "deeper" form of syntax or grammar, which is sometimes called the "logical form." Linguists have attempted to isolate these two forms of grammar so that they can be differentially studied. Articulatory grammar is more likely to be publicly taught, and learned, while the internal logical form apparently grows naturally "in the mind" (30). Internal mental representations are more likely to be based on the untaught language faculty (the I-language). For example, a house described based on external perceptual experience (4-bedroom, 3½-bath ranch, frame house with attached two-car garage, on 37 fenced acres with view of city lights) is very different from the dream-based logic description of the same house (circular table set against the wall, with chevron designs incorporated into the chairs, set with black dishes streaked with gold – the room with a open door to another room from which light is shining).

Current attempts to achieve linguistic differentiation between these two systems are rudimentary but promising. For a child, the initial stage of language faculty incorporates the intuitive general principles of I-language structure. Such aspects include concepts of locational nature: actor, recipient of action, instrument, event, intention, and causation. As we age and enter the educational system, phonemic awareness, inter-actions, and expressions from the motor and perceptual systems are learned in order to achieve a semantic and phonetic interpretation of utterances (31).

Components of language grammar are likely to have genetic and neu-roanatomical correlates. This finding is potentially congruent with the concept of brain maps as proposed by Damasio, in which an individu-al's specific memories are tied to neuroanatomical brain sites and neu-ral interconnections (see Chapter 6). It has been suggested that that the word for "cat" in all languages could be tied to a discrete neural config-uration denoting the letter "C" in the brain. To this point, there is little evidence that there is a shared neuroanatomical site for such specific lin-guistic representations. Even if linguistic knowledge were this concrete, any brain-based representation of "C" still require attributed meanings and associations. The network of interconnections required to develop a global concept for "cat" would requires the recruitment of activity from large disparate areas and systems of the sensory and motor apparatus, as well as from the cortex (32).

This difference between perceptually based "surface" and "deep" logical grammar can be used to explore the neuroanatomical setting for these language systems. Since articulatory phonetic representations uti-lize a different grammatical system from that used to organize the non-perceptual language syntax such as that incorporated into dreaming, and since language characteristics are apparently tied to structural neuro-anatomy, at least two different neurocognitive linguistic systems are likely to be present in the human central nervous system. The brain sys-tems processing the articulatory components of language are dependent on and required for our interaction with the external environment. This system is more likely to use the same brain systems that we use to process external articulatory language input and our motor response. The logical I-language apparently utilizes a processing system that operates independently of the perceptual system. We use the logical I-language of dreaming to develop concepts of causation, instrument, event, and intention. In developing these aspects of grammar, it is probable that we utilize a different anatomy, set of neural connections, and electrophysiology from that which we use for the linguistics of waking perception.

SOAP OPERAS AND OTHER PARABLES

Narrative organizes experience by drawing together aspects of spatial, temporal, and causal relationships into story – a special pattern that represents and explains experiences. Narrative organizes a group of experiences, sentences, or pictures so that they have a beginning, a middle, and an end. The beginning, middle, and end are not discrete elements such as words, grammar, or sentences; rather, they are relationships among the elements that make up the story. Narratives can reflect or vary a narrative voice as well as alter the point of view. Genuine narratives can expand or foreshorten time, altering temporal perspective as needed based on plot requirements (33). Narrative is a way of globally organizing a set of relationships, experiences, and a sequence of actions within a progressively changing four-dimensional construct (34). This ability to form and present narrative story is, like dreaming, a basic tenet of what is required of us in being human (35).

Literary narrative forms can be classified as dramas, epics, or lyrics. Dramas are compositions using dialogue that are designed as if to be acted out on stage. They most often involve conflict or contrast of character. Epics include a heroic character arc that is dealt with at length in narrative style, with subjects presented most often as external to the author. Lyrics are compositions in the form of song, more likely to be an expression of the writer's own thoughts and feelings. Narratives can also be classified based on genre, whether mainstream fiction, non-fiction, science fiction, mystery, western, etc. Dream narratives most often present the episodic quality of our lives as lived. They are closest in structure to the epic genre that we call the soap opera – the continuing saga of our imaginary lives. In this way, our dreams imitate not only the content, but also the form of our waking experience (36).

Dream narratives describe the content and experience of our internal world. Narrative organization is a basic characteristic of dreaming. Dreamers will organize even the most extreme of dream-associated experiences, the parasomnias of night terrors, hypnogogic hallucinations, and sleep paralysis, into narrative stories. As noted in Chapter 6, REMS dreams are longer than the dreams from other sleep stages. If the dream is to be remembered or shared, longer dream narratives, such as those of REMS, require more applied organization and structure than do the shorter visual images from sleep onset, and the confused emotional outbursts from deep sleep.

The capacity to organize dreaming into narrative, while potentially a species characteristic, is not a capability that is common to everyone. In our study of non-dreamers, many of the individuals included in the rare-dreaming category (see Chapter 1) initially reported that they did not dream. At interview, a subset of these individuals reported that what they recalled on awakening from sleep was a color, an emotion, or an awareness that a dream had occurred during sleep. These dreams were not presented as narratives. It is unclear whether this difficulty in organizing their dreams into narrative had any effect on their waking lives. All had the ability to function in waking, and most had paying employment in society. It might be expected that individuals who find it difficult or are unable to organize their waking experience into narrative stories would have trouble communicating and functioning in society. Psychiatric tests that address storytelling, such as the Thematic Apperception Test (TAT) (parables) and the Rorschach test (ink blots), were developed based on perceived psychoanalytic constructs for psychiatric illness. These tests can be helpful for the psychiatrist attempting to discover whether a patient has difficulties with reality testing, as in a patient with delusions, or in personality disorders when patients have problems with judgment. Such tests have rarely, if ever, been used to address the exceedingly complex area of waking narrative organization. Yet, a "good story" is what is required for most psychiatric diagnoses. A patient is required to present a "story" that provides for the therapist the relevant behavior and thought characteristics that are attributable to a particular diagnosis. The therapist uses the format of a structured interview to obtain from the patient the relevant information to meet criteria for one or several of the many psychiatric diagnoses included in the diagnostic manual for the field – the *Diagnostic and Statistical Manual of Mental Disorders* (DSM), now in its fifth edition (37). In the days before DSM and the development of psychiatric diagnostic criteria, psychoanalysts used the initially presented patient dream (the manifest dream) to make diagnoses and prescribe treatment. Today, psychiatrists no longer use this psychoanalytic approach to diagnosis. The narrative story presented in the manifest dream is, however, still used in some modern approaches to psychotherapy (38). It remains unclear whether individuals who have difficulty organizing their dreams into narrative form also have similar organizational difficulties with waking thought and waking functioning, or with meeting criteria for psychiatric diagnosis.

PRETTY HORSES AND THE STRUCTURE OF STORY

Stories are often based on master plots that are told over and over. The narrative structures of fiction follow a consistent pattern: the hero, the goal or objective, the helper, and the opposing force or hinderer. Heroic myths, their protagonists and characters, are the life-and-death forces of most narratives (39). The repeating characters of these myths, such as the young hero, the wise old man or woman, the shape-shifter, and the shadowy antagonist, have psychoanalytic correlates with the figures that appear repeatedly in our dreams and fantasies. Freud analyzed both dreams and stories for the presence of recurrent myths involving parent and child, in particular the Oedipus complex, in which the child is in love with one parent while hating the other. He pointed out that character relationships and attachments are often based on fixations that develop during childhood when feelings for the parent of the opposite sex are not transferred to a sexual partner outside the family. According to Freud, the Oedipus complex is present within the literary structure of most stories (40).

Post-Freud, one of the most productive avenues for psychoanalytic perspectives has been in the area of literature and film. Today, writers, film-makers, and other creative artists consciously use psychoanalytic codes, structures, and characters in the attempt to give their stories the ring of psychoanalytic truth (41). Film and literature critics often review a text symbolically, reading not for conventional meaning but rather for the presence and meaning of embedded symbols. Most symbols and archetypes are visual, and these symbolic approaches have been easily adapted to film by screenwriters and cinematographers. Some purely symbolic films attempt to tell stories using visual symbols free of the plot-driven script. A famous example is *Spellbound* (1945), in which a major surrealist artist (Salvador Dali) was brought in to create mysterious visual symbols to be used as props and in set design (42). These symbols promise a privileged way of knowing for the psychoanalytic viewer trained to decipher the "code." Yet in this situation where everything is a symbol, even with great actors (Ingrid Bergman and Gregory Peck) under the masterly direction of Alfred Hitchcock, the story itself is somehow lost. *Spellbound* as a film is visually concrete, driven by a vague and inconsistent plot, with an underlying story that is unapparent to many viewers.

In this past century, literary and film criticism has been dominated by the psychoanalytic approaches developed by Jung, Freud, and their followers. Symbols, signs, and codes have proven useful for describing the basis, motive, and meaning of both waking and sleeping fictions. The manipulation of psychoanalytic signs and symbols has become part of the writer's craft. When the storyteller incorporates a symbolic structure, he or she is evoking a peculiar kind of experience and understanding that is purportedly shared by all dreamers.

These interpretative approaches are external systems of analysis imposed on literature, and some writers have rebelled, choosing to view the psychoanalytic perspective as a stifling structural paradigm. D. H. Lawrence argued that "The Freudian unconscious is the cellar in which the mind keeps its bastard spawn" (43). Susan Sontag notes, "Interpretation is the revenge of the intellect upon art. Even more. It is the revenge of the intellect upon the world. To interpret is to impoverish, to deplete the world – in order to set up a shadow world of 'meanings.' It is to turn the world into this world. This world! As if there were any other" (44). In their response, writers explored older pre-Freudian works, studying how symbols and dreams were used in narrative before the era of psychoanalysis. Shakespeare used the dreamscapes embedded in his plays to enhance the status of characters and to present their inner worlds. He attempted in his plays to bring us, as readers and viewers, to the point at which we can "dream her dream for ourselves" (45). Other writers have used the insertion of dreams into text as a literary technique to incorporate multiple points of view, different forms of the poetic "I," the disguise of meanings, and to allow for narrative transformations to occur (46). Writers developed techniques to free their work of psychoanalytic codes and symbolic meanings. Such techniques as Joyce's theory of epiphanies and Eliot's notion of objective correlative were used to insert poetic interstices in their work that are deliberately vague and unanchored in pre-established codes and meanings (47). Strange, inexplicable and intrusive events, gestures, and images are included in text within a context that is too weak to justify their presence. Often these interludes are inserted as dreams, and like dreams, these passages apparently reveal something else. That "else" is left for the reader to determine.

> That night he dreamt of horses in a field on a high plain where the spring rains had brought up the grass and the wildflowers out of the ground and the flowers ran all blue and yellow far as the eye could see and in the dream he was among the horses running and in the dream he himself could run with the horses and they coursed the young mares and fillies over the plain where their rich bay and their rich chestnut colors shone in the sun and the young colts ran with their dams and trampled down the flowers in a haze of pollen that hung in the sun like powdered gold and they ran he and the horses out along the high mesas where the ground resounded under their running hooves and they flowed and changed and ran and their manes and tails blew off of them like spume and there was nothing else at all in that high world and they moved all of them in a resonance that was like a music among them and they were none of them afraid horse nor colt nor mares and they ran in that resonance which is the world itself and which cannot be spoken but only praised. (McCarthy, 1992) (48)

This Cormac McCarthy passage, in symbolic mode, inserts a literary dream into the dialogue and plot of the presented text and takes the reader into realms of understanding of the text that are less amenable to reason (49). Such an insertion, such a dreamscape, can call into question the virtual reality of the plot-driven text, involving the reader with a presented story in which virtuality potentially becomes reality.

DREAMING AND THREE-ACT STORIES

Storytelling differs from dreaming in that it is consciously designed to entertain and interest the reader or viewer. The storyteller has a responsibility to attract and maintain a reader's attention with a plot that requires coherence, tension, and crisis. In Stewart Stern's long narrative nightmare (included in Chapter 7), the differences between dream and story are particularly apparent. This dream can be read symbolically based on Freudian sexuality and relationships or Jungian symbols. Such an applied interpretive approach can have value in explaining the presented signs and symbols, but such a symbolic reading avoids addressing the narrative structure of the dream. This is the dream of an expert screenwriter, and its narrative structure is of particular interest. While Stern has used images from this dream in developing scenes in screenplays, he would never write a screenplay constructed as loosely as the story of this dream. This dream narrative is presented without applied organization, grammar, or expectation of critique. Yet this dream has structural organization. The presence of these structures even in dream suggests that they are among the basic techniques that we use to organize thought.

In reading Stern's nightmare as a narrative, it is clear that the dream, like most narratives has a beginning, a middle, and an end. In fact, this dream appears to have a three-act structure in which each act has different functions in the story – Act 1: the scenes from his childhood in New York in which the timing, place, setting, characters, themes, and rules for the story universe are established, as well as the genre (memoir as opposed to soap opera); Act 2: a road story of his journey that leads him across America in his search for the house, allowing for the arc of character to develop; and Act 3: the culminating story of the film being shot in his house that comes to a climax of hospital beds and lost relationships tied together within the author's cycle of life and experience. It is possible that Stern's dream follows this structure because as a screenwriter, he is used to presenting stories with that format. It is also possible that in transference, in recording this interview, I applied my own structure to the dream report. But it is more likely that the three-act structure is the typical way in which we cognitively organize the narratives of longer stories and dreams. Freud's dream of Irma's injection (see Chapter 2) is structured in a similar manner – 1: the party and the presentation of symptoms (timing, place, setting, characters, themes, and rules for the story); 2: the physician's visit, interactions, and treatment (the journey); and 3: climax, conclusion, and judgment.

The other long, likely REMS-associated dreams included in this text also have this structure. Mary Shelley's nightmare of *Frankenstein* has an Act 1, in which the monster is brought to life; an Act 2, in which the pale student of unhallowed arts attempts to escape from his act through sleep; and an Act 3, in which the hideous corpse returns to his bedside and examines him with yellow, watery, but speculative eyes. My shorter, REMS behavior disorder, bear dream (see Chapter 6) has an Act 1: writing on the porch; Act 2: flight to escape the bear; and Act 3: confrontation and actions on coming in contact with the bear.

The shorter sleep-onset dreams of intense visual imagery (see Chapter 6) and the deep-sleep dream of rolling stones (see Chapter 1) have no such structure, nor does the longer but less developed stage 2 anxiety dream of a number (see Chapter 6). It seems evident that the length of the narrative report and the sleep stage of origin are likely key determinants as to whether a dream narrative is presented with a beginning, a middle, and an end – a three-act structure.

Stewart Stern's nightmare is not a myth. There is only a limited heroic interplay between the dream characters. There is the suggestion, however, that such a mythic structure could develop. The first person dreamer could be a hero, the realization of returning to his home to resolve unfulfilled relationships is the goal/objective, Joanne Woodward and Beatrice Lillie are potential helpers, and Heath Ledger is the opposing force or hinderer. There is a plethora of signs and symbols. This dream in the hands of an artist such as Stern could be inspiration for the screenplay of *On the Waterfront 2 – Aftermath of the Battle*. However, any heroic and/or mythic structure that is present in the dream is incompletely developed. Some of that structure could come from the transcriber and even the reader who is culturally trained to approach story in such a manner. In writing down this nightmare and adapting the unresolved trauma and horror of the dream to narrative, I brought to the story my tendency to view as heroes veterans able to overcome their trauma. It is clear that, beyond Stern's report of his dream, I contributed as transcriber to the construction of a coherent and heroic story that was suitable for presentation. Such a heroic/mythical structure is less clear for the other dreams included in this text, though Irma's injection has been considered as such, in part, based on the writer and interpreter's tendency to either deify or demonize Freud (50).

A primary requirement for the development of a character in film or literature is for the character to change in an arc across the story so that the presented experience has an effect that is mirrored in the development of the character. There is only a limited character arc in Stern's nightmare: the change in age and location, and the realization of unresolved loss. Lack of resolution is typical of a nightmare, and it may be that limited character development is characteristic of post-traumatic stress disorder (PTSD) nightmares, in which the dream is repeated, stereotypic, and without resolution of the initial trauma. Again, however, the recording and writing of the dream had its own dynamic. When Stern read this version of his nightmare, the emotion, disgust, horror, and helplessness associated with his nightmare did not change. However, when he read the transcription and approached his nightmare as an externally presented story, his memory of the long-ago trauma changed. More than sixty years after the Battle of the Bulge, he now remembered the actual history of the traumatic event rather than the emotional angst of the experience in a way that made him wonder what had been upsetting him so for so long. In the two years since the interview in the Zen garden, his nightmare has not returned.

DREAMING AS A STRUCTURING PRINCIPLE FOR LIFE

Dreams offer an available construct of emotion, memories, images, and alternative logic of such value that they are commonly used by artists in their creative process. Great stories sometimes originate as dreams. The close reading and assessment of the formal structure of Stewart Stern's nightmare suggest that Harry Hunt was correct in his realization that dreams typically fall a bit short in attempting to become genuine stories (51). However, it is interesting how close they can come, particularly when reported by a master storyteller such as Stern or Freud.

Freud and Jung are gone but their followers have persisted in pointing out that our dreams are filled with signs and symbols that tell a story that is far different from the catoptric, plot-driven, and apparent narrative that we report as a dream. These signs and symbols can be used by the dreamer as an approach for helping to understand the meaning of a dream. They can be used by writers and film-makers to develop the internal world of a character, and create a more believable, virtual impression of reality for the reader and viewer.

The signs and symbols of dreams are the building blocks that are used to construct stories with a narrative structure that is present in dreams in their most basic form. Linguistic theory and research suggest that the mentation of dreams is organized using an internal (non-perceptual) and logic-based grammar that differs from that used to organize external waking experience. The writer uses this logical grammar to construct plot, actions, and character. This logical construct is the template on which dialogue and visual descriptions are built. The writer, in using this shared organization and construct of the I-language, can bring the reader or viewer into an inner world that has structural characteristics very similar to his or her own.

Dreams have typical narrative structures that resemble most closely the epic form of story we call the soap opera. Longer dreams and those from REMS are those most likely to have a three-act structure, a structure that forms a basic organizing pattern for story, and perhaps for thought. Myths and heroic forms are shared culturally, perhaps archetypically, and are characteristic of both dreams and stories. The presence or absence of character arc in a dream may, in part, differentiate the content of a typical dream from that of a typical PTSD nightmare. The signs, symbols, grammars, and narrative structures apparent in dreams and stories describe our attempts at comprehension of our world. It may be that this structuring reflects a basic tenet of what is required of us in being human, "learning to understand and to be able to tell stories" (52). As Hunt (1991) points out, "the creative tension required to produce novel, emergent forms of self knowledge requires a staged collision between subjectivity and objectivity that may characterize the dream" (53). This process, available to every dreamer, is what the writer tries to recreate. The relevance of a dream's content to a dreamer's personal life may be less important than the function that dreaming serves as a structuring principle for that life.

The writer can use these dream-based techniques of symbology, I-grammar, and myth-based three-act structure to create for the reader a story that resembles a dream. The writer can also use these techniques to create an experience for the reader that looks, sounds, or feels as if it were occurring in the exterior and perceptual world. This is possible, in part, because beyond their content of thought, emotion, and memories, dreams are most commonly experienced as a series of visual images. And in those visions, whether based on waking experience or on inner imagery, we have the opportunity to virtually integrate and organize our perceptual experience of the exterior world.

Cormac McCarthy's literary dream of horses is the exception. It is a longer dream that attempts, as far as possible, to avoid organization or structuring. That is because, although described as a dream, it is not written as a dream is normally experienced. This literary dream is happily inserted in the midst of a dark story, of loss, death, and being trapped without hope – a story that is an experiential nightmare. There is no three-act structure. There is no myth or heroic arc. There is only the vision of pretty horses running in an almost musical resonance across the prairie through translucent pollen and the powdered gold of the sun. The emotions are positive; in most dreams they are negative. This literary dream is even more timeless and unstructured than an experienced dream. And yet it seems very much like a dream, more dream-like than many of the actual dream transcripts (Stewart Stern's included) that are presented in this text. In the next chapters, we will explore this visual stage of dreaming and how an author or a film-maker might accomplish such a feat.

Notes

1. Poe, E. A. (1849) "A Dream Within a Dream," first published March 31, 1849, in *Flag of Our Union*. This famous poem/quotation resonates in a precognitive voiceover at the start of Peter Wier's iconic Australian film, *Picnic at Hanging Rock* (1975).
2. Garfield, P. (2010) *The Universal Dream Key*. New York: Harper Collins.
3. States, B. (1994) Authorship in dreams and fictions. *Dreaming* 4: 237–253.
4. Hunt, H. (1991) Dreams as literature/science: An essay. *Dreaming* 1: 235–242.
5. Koestler, A. (1969) *The Act of Creation*. New York: Macmillan, p. 45.
6. Fruman, N. (1972) *Coleridge: The Damaged Archangel*. London: Allen and Unwin, pp. 335–338.
7. Van de Castle, R. (1994) *Our Dreaming Mind*. New York: Ballantine Books, pp. 12–21; Pagel, J. F. (2008) *The Limits of Dream – A Scientific Exploration of the Mind/Brain Interface*. Oxford: Academic Press/Elsevier, pp. 156–158.
8. Shelley, M. (1818) *Frankenstein*, in Grant, J. (1984) *Dreamers*. Bath: Ashgrove Press, pp. 82–83.
9. States, B. O. (1993) Bizarreness in dreams and fictions, in *The Dream and the Text*, ed. C. S. Rupprecht. Albany, NY: SUNY Press.
10. Chomsky, N. (2000) *New Horizons in the Study of Language and Mind*. Cambridge: Cambridge University Press, p. 5.
11. In Aristotle's *Rhetoric* as interpreted by Eco, U. (1984) *Semiotics and the Philosophy of Language*. Bloomington, IN: Indiana University Press, p. 28.
12. McGrath, T. (1982) Language, power and dream, in *Claims for Poetry*, ed. D. Hall. Ann Arbor, MI: University of Michigan Press, pp. 286–295.
13. In Aristotle's *Rhetoric* as interpreted by Eco, U. (1984) *Semiotics and the Philosophy of Language*.
14. Eco, U. (1984) *Semiotics and the Philosophy of Language*. Bloomington, IN: Indiana University Press, p. 3.
15. Pierce, C. (1931–1958) *Collected Papers*. Cambridge, MA: Harvard University Press, p. 228.
16. Hobbes, T. (1651/2013) *Leviathan, Printed for Andrew Crooke at the Green Dragon in St. Paul's Churchyard*. A Public Domain Book, p. 3.
17. de Saussure, F. (1959/2011) *Course in General Linguistics*, trans. W. Baskin, ed. P. Meisel and H. Saossy. New York: Columbia University Press.
18. Eco, U. (1984) *Semiotics and the Philosophy of Language*, pp. 21, 44.
19. Freud, S. (1899/1965) *The Interpretation of Dreams*, trans. and ed. J. Strachey. New York: Avon Press, p. 351.
20. Jung, C. (1948/1974) General aspects of dream physiology, in *Dreams*, trans. R. Hull. Princeton, NJ: Princeton University Press, p. 52.
21. Jung, C. (1948/1974) On the nature of dreams, in *Dreams*, trans. R. Hull. Princeton, NJ: Princeton University Press, p. 68.
22. Jung, C. (2009) in *The Red Book: Liber Novus*, ed. S. Shamdasani. New York: W. W. Norton & Co. Among psychoanalysts, Jung has for me always been the easiest to approach and read. Where Freud is sometimes opaque and self-referential and Lacan coded, Jung is the most logical and clear in his descriptions. And then there is the *Red*

Book. Was this a difficult experiment, a heroic journey, a broken myth, or is this evidence of a schizophrenic episode suffered by one of the fathers of psychoanalysis? The *Red Book* is more likely to be a marker of Jung's eccentricity than his psychiatric dysfunction. Viewed retrospectively, based on literature and historical analysis, most of the writers of our major religious texts would meet DSM-IV criteria for major psychiatric diagnoses and perhaps long-term care.

23. Wolfson, E. (2011) *A Dream Interpreted Within a Dream: Oneiropoiesis and the Prism of Imagination* (uncorrected proofs). Brooklyn, NY: Zone Press. It is amazing what you can sometimes find in a used book store.

24. *American College Dictionary* (1947). New York: Random House, p. 366.

25. Pagel, J. (1999) A dream can be gazpacho. *Dreamtime* 16: 6–8; Pagel, J. (1999) Proposing a definition for dream. *ASDA Dream Section Newsletter.*

26. Pagel J. F. (Chair), Blagrove, M., Levin, R., States, B., Stickgold, B. and White, S. (2001) Defining dreaming – A paradigm for comparing disciplinary specific definitions of dream. *Dreaming* 11: 195–202.

27. Domhoff, W. (2003) *The Scientific Study of Dreams: Neural Networks, Cognitive Development and Content Analysis.* Washington, DC: American Psychological Association.

28. Eco, U. (1984) *Semiotics and the Philosophy of Language*, p. 14.

29. Chomsky, N. (2000) *New Horizons in the Study of Language and Mind*, pp. vii, ix, 5–9.

30. Chomsky, N. (1981) *Lectures on Government and Binding*. Dordrecht: Floris; Chomsky, N. (1986) *Knowledge of Language*. New York: Praeger.

31. Chomsky, N. (2000) *New Horizons in the Study of Language and Mind*, pp. 45, 60, 62.

32. Fodor, J. (1975) *The Language of Thought*. New York: Crowell.

33. Ricoeur, P. (1984) *Time and Narrative* Vol. 1. Chicago, IL: University of Chicago Press, p. 150.

34. Brannigan, E. (1992) *Narrative Comprehension and Film*. London: Routledge, p. 3.

35. Foulkes, D. (1985) *Dreaming: A Cognitive–Psychological Analysis*. Hillsdale, NJ: Lawrence Erlbaum Associates.

36. States, B. (1997) *Seeing in the Dark: Reflections on Dreams and Dreaming*. New Haven, CT: Yale University Press, p. 215.

37. DSM-V, *Diagnostic and Statistical Manual of Mental Disorders* 5th ed.) (2013) Arlington, VA: American Psychiatric Association.

38. Siegel, A. (2010) Dream interpretation in clinical practice: A century after Freud, in *Dreaming and Nightmares*, ed. J. F. Pagel, *Sleep Medicine Clinics* 5(2). Philadelphia, PA: Saunders/Elsevier, pp. 299–315.

39. Todorov, T. (1981) *Introduction to Poetics*, trans. R. Howard. Minneapolis, MN: University of Minnesota Press.

40. Freud, S. (1899) *The Interpretation of Dreams*, pp. 294–297.

41. Vogler, C. (1992) *The Writer's Journey: Mythic Structure for Storytellers and Screenwriters.* Studio City, CA: Michael Wiese Productions.

42. MacPhail, A. and Hecht, B (1945) *Spellbound* [Film], an adaptation of the novel *The House of Dr. Edwardus* (1927), H. Saunder and J. Palmer. British Board of Film Classification 1946-01-30.

43. Lawrence, D. H. (1921/2005) *Psychoanalysis and the Unconscious*. Mineola, NY: Dover, p. 9.

44. Sontag, S. (1966/2001) *Against Interpretation, and Other Essays*. New York: Picador.

45. Holland, N. (1993) Hermia's dream, in *The Dream and the Text*, ed. C. Rupperecht. Albany, NY: SUNY Press, p. 197.

46. Goldberg, H. (1993) The Marques de Santillana: Master dreamer, in *The Dream and the Text*, ed. C. Rupperecht. Albany, NY: SUNY Press, p. 245.

47. Eco, U. (1984) *Semiotics and the Philosophy of Language*, p. 157.

48. McCarthy, C. (1992) *All the Pretty Horses*. New York: Alfred A. Knopf, pp. 161–162.
49. Lee, C. (1993) The prophetic dream, in *The Dream and the Text*, ed. C. Rupperecht. Albany, NY: SUNY Press, p. 302.
50. Kramer, M. (2007) *The Dream Experience: A Systemic Exploration*. New York: Routledge; Thornton, E. M. (1983) *The Freudian Fallacy*. London: Blond & Briggs.
51. Hunt, H. (1989) *The Multiplicity of Dreams: Memory, Imagination and Consciousness*. New Haven, CT: Yale University Press.
52. Foulkes, D. (1985) *Dreaming: A Cognitive–Psychological Analysis*.
53. Hunt, H. (1991). Harry is the pro at writing such short and beautifully impactful descriptions.

Visual Dream Consciousness

Art thou not, fatal vision, sensible
  To feeling as to sight? Or art thou but
  A dagger of the mind, a false creation
  Proceeding from the heat-oppressed brain? **(Shakespeare,** Macbeth **) (1)**

Humans, like most animals, are visual beings. We use vision as our primary sensory system for understanding the external environment. Our vision presents the external environment to us in a comprehensible manner, allowing us to maintain a rational interaction with the reality outside ourselves. We take the perceptual information available to us from complex visual sensors (our eyes), and neurologically process that information into a representation of the external universe that we can interact with and manipulate on a consistent basis. This perceptual construct of external reality is but one of many possible representations. In our immediate biosphere there are organisms that operate successfully in very different sensory environments: the dog has an extended aural and olfactory range; birds have differences in color vision, acuity, and extended frequency range; and fish demonstrate electromagnetic sensitivity and a visual system designed to function in a very different medium from air. Some mammals (cetaceans) and birds (penguins and

163

swifts) can shut down the vision in half of their brain and sleep in that portion while they continue to swim or fly using the other half.

There are a multiplicity of organisms utilizing chemical, electrical, and substrate sensors that provide a perspective of the exterior universe that is unbelievably different from ours. And these organisms are using our three-dimensional concept of external space. Mathematically, three-dimensional space is only one among many of the possible representations available. Added dimensions potentially can be used to describe external space even better than our three-dimensional coding system. Yet we function fairly well in this universe utilizing our structurally and cognitively limited visual representation of external reality.

The perceptual isolation defining sleep consists primarily of a lack of vision. Sounds, pain, and touch are sometimes incorporated into dreams, but not the sensory experience of sight while we are asleep. We create our dreams from memories, emotions, and thoughts that we organize into stories. Yet these components of dreaming are most often experienced as secondary to the remarkable visual imagery present in our dreams. Images can have exceptional power to affect our waking thoughts and behavior. "Contextual images," whether the falling of buildings or the undertow of a giant tidal wave, can overwhelm us in our dreams (2). These powerful visuals can be archetypes, the shared, mystical, indefinable, and infinitely interpretable symbols that occur during dreaming (3). They can serve as the basis of inspiration and story, the central context around which everything else forms in the dream or in the story.

Both waking perceptual vision and imagery (non-sensory visual constructs) utilize a cognitive processing system that is based on shared patterns of representational images. These representations have a far more mundane and important function than their proposed role in providing psychoanalytic insight into the inner functioning of the mind. We use these images as a system of internal coding in order to make sense of the external perceptual world. Our intrinsic memory system includes the codes for object forms. When we look at an object and describe it as a "chair," that perception is based on our intrinsic memory for chair – the attributes that apply to many objects viewed from different perspectives that are all chairs (4). These are shared codes, so that what you know as a chair is also what I see as a chair. Intrinsic representations, sometimes called "representational primitives," do not necessarily need to be learned, and are probably part of our biological (genetic) endowment. In the central nervous system (CNS) we use these representations to integrate the external images of perception with higher order cognitive systems. We each have a wide variety of these basic codes of representation that we use to categorize viewed objects. We also use such representational primitives to categorize surfaces and events (the temporal analogues for objects). Each of these categories of "primitive" has specific parameters, relationships, and transformations that govern its relation to other primitives. For instance, surface has parameters of color, depth, texture, and orientation, as well as relationship aspects of junction, edges, concavities, convexities, gaps, and holes (5). Visual sensory input provides the triggering clues that activate the specific surface, object, and event codes that we utilize in categorizing, processing, and theoretically grouping perceptual phenomena. These representational primitives form an internal "I-language" of visual processing.

Visually, the exterior world with which we interact has a remarkable level of perceptual complexity. We respond to that complexity by triggering multiple and sometimes overlapping internal representations of these intrinsic perceptual primitives. In order for us to make sense of an environment that includes multiple (conjoint) pictorial representations, our visual system uses additional mechanistic approaches to define the relationships between primitives. To make sense of a complex and sometimes confusing environment, we use internal computations for characteristics such as space and depth, color, brightness, and shadow, to structure the relationships between multiple conjoint primitives in a visual field. These constructs form a visual I-grammar/syntax that we use to structure the relationships between conjoint representations of objects, surfaces, and events.

The imagery of dreaming occurs during sleep, a conscious state of perceptual isolation, and only a portion of this complex sensory visual integration system is needed for cognitive processing. When we are interacting with the exterior world or attempting to create representations of that world, we require continuous, explicit, and accessible perceptual input in order to describe the objects, the relationships between shapes, color, texture, and the spatial relationships between each point in the visual field. Sensory visual processing is driven primarily by this sensory input. Dream imagery is, however, derived from representations stored in memory (6). Non-perceptual images, such as those in dreams, are not fully depictive, and do not necessarily have an arbitrary relationship to the thing represented. Nor are these images the simple re-embodiments of stored sensations.

Since the visual images in dreams are often what we remember, the limitations of dream imagery may not be consciously apparent. This thought space of dream imagery is very different from the visual perception of sensory space (7). Dream images, although they appear to be images from the real world, are actually two-dimensional, appearing as three-dimensional space. Perceptions based on images from the external world have a vitality and vivacity that such non-perceptual imagery often lacks (8). As Jean-Paul Sartre observed in 1940, if the object we select to imagine is the face of a close friend, one known in intricate detail, it will be, by comparison with an actual face, "thin," "dry," "two-dimensional," and "inert" (9). Eileen Scarry likes to ask attendees at her writing workshops to visualize a rose with their eyes closed. After several minutes she asks them to open their eyes and look at an actual rose (10). The difference is remarkable. Relative to the actual rose, the virtual rose of mental imagery is but a pale, simplistic, and translucent reflection.

The virtual (prepositional) system with its representational I-language and grammar is an extremely important component of CNS functioning. A majority of waking sensory input is processed by this system (11). It is possible that non-perceptual imagery uses propositional representations functions independently of the "depictive" systems that we use in sensory processing in order to cognitively organize the exterior world. It is far more likely, however, that imagery and visual sensory perception share underlying systems of neural processing (12). So, in order to better understand the visual systems active in dream imagery, it becomes necessary to explore, in at least a superficial way, some of the recent scientific progress in the complex area of visual processing.

ARCHETYPICAL VISUAL PROCESSING

The major requirement for any perceptual system is consistency, and we are able to use our visual systems to present the external environment to us in a comprehensible manner. But this is not a simple process. Visual processing requires the largest area of the neuroanatomical CNS of any cognitive function. In the monkey, there are at least thirty-two discrete areas of the cerebral cortex that respond directly to selective visual input (13).

The eye is a remarkably complex sensory organ. Some authors argue that the eye is a visuospatial system with the ability to approximate consciousness (14). Retinal cells are responsive to a wide variety of stimuli ranging from color to brightness, darkness, orientation, and motion (15). Data derived from these receptor cells are processed by multiple interactive parts of the CNS that process and evaluate visual input. There are at least two major sophisticated cortical systems that function to integrate visual content with memory systems and conscious activity (16). One of these systems is organized based on point of view, the description of aspects of the visual field relative to the viewer. Another major system functions to analyze the relationship of one object to another in the visual field independent of the observer (17). Each of these systems is strongly interconnected, existing in parallel in each cerebral hemisphere and processing the information from the contralateral eye.

There is also a major subcortical system that functions at a nonconscious level to control eye movements and cross-modal spatial processes. The soldier in war, the tennis player reacting to a serve, and the engrossed videogame player can respond to changes in the visual field at speeds faster than those required for conscious mental processing. The baseball player reacts to a pitched ball with extraordinary non-conscious speed; bracketing possible trajectories, organizing complex muscle groups, addressing alternative possibilities and responses, and planning for the next actions in a seamless process that can have aesthetic beauty. This independently functioning subcortical system is sometimes called the "visual buffer." This system routinely operates without conscious controls.

Our understanding of visual cognition has changed markedly in the past few years based on the realization that many CNS visual processing areas are organized to depict visual patterns of retinal reception as projected on to the cells of the brain. Approximately half of the visual processing areas in the brain are topographically organized so as to represent externally perceived space and objects (18). The strongest study demonstrating this process was conducted using a monkey who had been trained to stare at a geometric pattern of flashing lights. The monkey was killed after being injected with a radioactive sugar that is taken up by nerve cells based on their level of activity. In the monkey's occipital cortex (a primary visual area of the CNS), the same spatial pattern at which the monkey had been staring was found, inscribed as a pattern of cell activation (19). Such studies indicate that sensory visual data are neuroanatomically presented as topographically based neural input. Within the neural architecture of the brain a map of sensory visual input is constructed onto networks of nerve cells. This internal display approximates the visual form of an object presented in analogue format – an analogous model with components of correspondence and similarities to the visually perceived object. This system uses presented visual data to create a computationally described representation of that object in the neuroanatomy of the brain. In other words, visual images actually exist in the brain as neural images.

The images processed through this system have different attributes and characteristics from sensory-based images. This "propositional" mental imagery is different in several basic ways: it is slower and more controlled, without the stimulus-based attention shifting that typifies external perceptual input; imaged objects fade quickly while external objects persist as long as they are present; images are limited by the information encoded into memory; and we can exert control over objects in images when the external world is most often outside our volitional control (20). Depictive representations of the external environment include all aspects of shape, relationships to shape, color, texture, and spatial relationships in the visual field. The propositional mental representations are abstract, often referring to classes of objects rather than a specific object. Mental images are fleeting. The input to topographically organized areas changes every time our eyes move so that the projected pattern is lost. These mental images are not simple re-embodiments of stored sensory data. This is information organized in a fashion that can be utilized by the CNS in cognitive processing. A visual object is cognitively processed based on its mental construct, with each representation having cognitive information associated with that particular mental image. There are aspects of higher consciousness attached to such images; these can include an individual's associated memories, emotions, beliefs, attributions, and expectations. This cognitive penetration can alter an image so that the image is processed in collateral cognitive processing systems. As an example, emotion attached to an image can qualify that image as an emotionally competent stimulus able to affect the visual buffer, emotions, feelings, and even the dreams in which these images are combined and altered to produce unexpected emergent forms (21).

Vision is among the most neuroanatomical of cognitive processes. Positron emission tomography (PET), functional magnetic resonance imaging (fMRI), and other scanning modalities indicate that there are demarked sites in the CNS used for the various cognitive processes involved in visual perception. Neuropathological damage to specific areas results in consistent patterns of visual defects. There is a close neuroanatomical correlation with cognitive process, more than for most other areas of cognitive neuroscience, and as a result, the visual system has been particularly available to modern technology. The advent of each new scanning technology leads to marked advances in this field. Many neuroscientists now conduct their work in the area of visual cognition, an area in which significant progress is being made. This perceived close association between neuroanatomy and visual cognitive processing has altered the field of cognitive neuroscience. Neuroscientists with a vision-based perspective tend to view all areas of cognition (including dreams) as neuroanatomical processes.

It is very likely that the mental imagery in dreams is also processed through this system. This topographically projected system is in many ways analogous to the concept of semiotic signs: "A sign is a picture if the perception of the essential properties that the sign has in relevant respects is identical to the perception one would have of the corresponding properties of some other object under a certain perspective and if this perception is constitutive for the interpretation of the sign" (22). There is excellent evidence that the shared representational constructs that we use to process visual information exist virtually as visual constructs in our brains. This is based on the known association of visual processing with neuroanatomy, what we know of the topographic prepositional representation of objects in the visual cortex, and our ability to format nonperceptual imagery using an operative cascade of visual processing (see Figure 1.1 in Chapter 1).

The scientific establishment has remarkable capacities for engineering technology. Biological processes once concretely described can be artificially created. Today, non-perceptual imagery processing can be incorporated into the visual systems of projected film and artificial intelligence. If he were still with us, Carl Jung would be amazed to find that over time his concept of shared archetypes/primitives has become less relevant when applied as a description of the structure and functioning of the mind, and far more useful when used as an explanation for the cognitive structures of visual imagery, the same systems that have been used to develop sensory visual processing systems for artificial intelligence (AI).

FOOLING AND BINDING VISUAL SYSTEMS

Much of what we understand about the processing of visual information comes from studying techniques that fool us into believing that we are seeing something that we are not. Vision is replete with various illusions, both physical and psychological, as well as visual phenomena such as ambiguities (visual projections that occasionally match object reality), paradoxes (those that cannot be matched or measured as objects), and fictions (that have no objective counterparts). There are techniques that can be used to manipulate and fool the sensory organ (the eye), the digital visual processing systems, and the propositional (topographic) systems (23).

Film-makers fool the "eye as the camera" into perceiving a continuous moving picture that has three-dimensional characteristics of depth and foreground by using a projected series of still two-dimensional images. In cinema a series of still photographs is projected at a speed of sixteen to twenty-four frames per second, and perceived as continuous. The periodic repetition of pictures dissolves perceptually when presented at a rate beyond this threshold into a continuity that is the mechanistic basis for the motion picture. Television presents at twenty-five pictures per second, each given twice to raise the "flicker" rate to fifty per second. A light flashing at a rate greater than fifty flashes per second appears to be an unwavering steady light. At these speeds of information presentation, our visual system cannot process the external information any faster than it can transmit it internally. The gaps between pictures catch our attention only when they are long enough for us to perceive the duration of the gap separately from the surrounding images. While early films, shot at sixteen or eighteen frames per second, are perceived to flicker based on their slow and uneven speeds, the perception of continuous motion requires even fewer images than the continuous projected photographs of the movie camera (24). We can recognize a two-dimensional visual depiction of an object that is presented in as little as 100 milliseconds (25).

The projection of images at a fixed rate creates a subordination of time to movement. In chronological time, the present is a point that moves continuously from the past into the future. A series of photographic images can be used to establish for the spectator what amounts to an almost complete disconnection with objective reality. There is a fundamental movement of time that has occurred and cannot be remade. Time is dynamic, a process of action that defines the adjacent and contiguous spaces of the picture. Each shot is linked in sequence to the next, each sequence into parts, and the parts into the whole of the film. An alternative reality is created that can be visualized outside the chronological reality of our external (non-film-watching) lives (26).

The subconscious "visual buffer" system can also be externally manipulated, inducing us to focus our attention on part of an image or lead us to disengage from a representation. The most powerful triggers for this system are the emotionally competent stimuli that bypass the normal visual processing system and divert attention to those signals even when we are not paying close attention (27). Emotional stimuli can change our focus and attention to a specific part of the visual field or to a visual representation stored in associative memory that was not part of the initial image.

Film-makers have discovered other ways to manipulate conscious visual processing. A repetitive presentation of still images is most often perceived as continuous (a visual characteristic occurring secondary to inherent time-based limitations in sensory–neural interactions). Cohesive objects are perceived as capable of tracing only one potential path through both space and time. We live in a constantly changing universe and tend to view objects and forms as being in motion (28). We visually anticipate the visual consequences of actions and events. Since in our exterior sensory world, objects are rarely visualized in the static condition, images project along a potential trajectory before that sequence is ever shown. In a constructed film scene, the viewer follows characters and objects along an arc of motion even when obstructed from view. In presented interactions between characters, the viewer follows the eyeline of contact. We visually extrapolate progressions of actions to final results. If we start to visualize what we are seeing as a succession of disconnected images, our attention wanders (29).

Motion can include changes in both color and form. The brain registers the change in color first, before registering the change in direction of motion. This difference is not one of perception, but one of the different processing speed required for each of these cognitive systems to integrate perceptual input (30). The sensory visual cognitive process can also be manipulated in the way in which we perceive physical space. Renaissance artists learned to present spatial elements in a picture by mimicking the exact geometrical relationship of how light would meet the eye in a three-dimensional scene. This created a consistent system so that a realistic appearance in depth and space could be achieved by the artist for the viewer (31). Perspective became the manner in which three-dimensional space was represented on a two-dimensional canvas. This system of artistic construction rules for portraying space and depth can be used to explain visual optics and the formation of images in the eye. Visual presentation in photography, art, or film is most often viewed and judged based on the effects that changes in viewing angle, height, motion, and distance have on the viewer's perspective. The system of linear perspective provides a consistent approach that translates three-dimensional space into two-dimensional representations in a way that we can all understand (32).

Fighting against the constraints of linear perspective, modern artists have developed alternative schemes for depicting pictorial space. Linear perspective can be viewed as culturally based, a projection of Western patterns for organizing the visual world (33). The convention of visual perspective may be nothing more than a highly selected stereotype that differs from those adopted by other cultures. Today, the idea that there is only one way to represent space in artistic renderings and photographs is often considered to be a matter of cultural bias or chauvinism (34). Studies from anthropology call into question the neuroscientific theories that have postulated a neurological basis for the rules of perspective. Lived experience and cultural context also affect the consistency of our visual worldview, and can alter the patterns of functioning in the human CNS.

Perspective as a component of visual representation is present in both dreams and imagery. Perspective may be a derivative construct of the process of topographic representation in the brain. Perspective provides a computationally useful approach to the representation of images and their relationship to other images in space. We can use the system of linear perspective to quantify and compare the size, orientation, and relationships between objects, consistently modeling and codifying our relationship to external sensory space.

The topographic processing system can be visually manipulated in other ways. The cave artists discovered that lines and surfaces of an object could be used to create a "primitive" representation of the object. They created images that the viewers "see" to be that object, differentiating between objects as diverse and similar as a mastodon, a horse, or a man and woman. In these images the visual representation of an object is achieved with either a global shape or part of a shape that can represent the particular object. The cave art renderings are free floating in space. Multipart images utilize spatial relationships between these representations of forms to create an image. Today, we understand that the visual continuity of such complex objects is maintained neurologically by relationships that we describe as "binding." There are at least seven different types:

- Property binding – different properties such as shape, color, and motion are bound to the objects that they characterize.

- Part binding – the parts of an object are segregated from the background and bound together.
- Range binding – particular values of a property such as color are defined within the dimension of that property.
- Hierarchical binding – the features of shape-defining boundaries are bound to the surface-defining properties of that object.
- Conditional binding – the interpretation of one property (e.g., motion) depends on another (e.g., depth, occlusion, or transparency).
- Temporal binding – successive states of the same object are integrated across temporal intervals as in real or apparent motion.
- Location binding – objects are bound to their locations (35).

All of these binding patterns were incorporated by the artists into the cave paintings. The pictures are independent, segregated from the background, with colors ascribed to individual paintings. There is an interplay between shape and cave surface. Objects overlie one another, maintaining congruency, and appear to move in temporal units, with apparent motion created by the repetitive depiction of the same creatures. The art is bound and protected by its location deep in the earth.

We use these techniques of visual binding in "seeing in" to observe objects. Sometimes these objects were never intended to be or actually believed to be representations. We see objects in stained walls and in relationships between disparate, overlapping shapes in our visual field. We see objects in Rorschach blots. It has been suggested that this is a potential origin for the cave art, with the artist accentuating shapes and protrusions in the cave walls with charcoal and colors so that in flickering artificial light those marks might appear to be the animals that the artist perceived in those shapes and shadows (36).

THE SPROCKET IN THE PROJECTOR

Dreams are most often remembered as images. And these images differ markedly from the images of waking perception. Dream images resemble those from the waking state of eyes-closed imagery, not generally incorporating perceptual input, particularly visual stimuli from the external environment. Dream images, except in the special case of lucid dreaming, also differ from waking imagery in that they are independent of our volitional conscious control.

The images of dreams and imagery are different markedly from those of waking perception in their independence from the sensory component of the visual system. These images are neither depictive nor simple re-embodiments of stored sensations. Images are fleeting, changing quickly and easily lost to waking conscious recall. Dreaming images and waking imagery rarely include real-time perception, emphasizing the neurobio-logical basis for the time-associated visual characteristics of seeing (37). The subordination of time to movement, the present as a point moving continuously from the past into the future, and the ability to perceive changes in chronological time require an available physiological marker. In waking visual perception, this marker is most often the set physiologi-cal time-lag required for the transmission of sensory information from the eye to the neural areas of cognitive processing. Our perspective of change in chronological time is based at least in part on the set limits of this physiological marker. In dreams, the dreamer often moves back and forth on a less defined timeline of associative memories and experience. Chronological time may not be present in a dream.

Physiological processes most often have functions. Sometimes these functions (e.g., the extracellular electrical fields of the CNS) await expla-nation or the development of scientific theories able to incorporate alternative perspectives. Since all physiological processes require an investment of metabolic and genetic effort, non-functional physiologi-cal systems are rarely, if ever, preserved. It has been proposed that the eye movements present during rapid eye movement sleep (REMS) are a reflection of dream cognition (e.g., watching a tennis match during dreaming) (38). In the search for physiological function, it has been pro-posed that eye movements during sleep are necessary to maintain corneal heath (wetting of the epithelial surface) in the eye during sleep (39). Eye movements occur most often during REMS, the sleep state that is most likely to include long narrative dreams (see Chapter 6). Narrative is most easily structured around a plot that moves from past to future. Repetitive rapid eye movements occur at set frequencies. There is a set time limit for the motor neural conduction that results in these eye movements. These eye movements have the potential to function in providing a time-based marker that is most active during the long dreams of REMS, utilizing a neural time-lag similar to that active during waking visual perception. Rapid eye movements could potentially function as the sprocket in the projector, adding the fourth dimension of time to dreaming.

Dream images, existing independently of the sensory modality of the eye, are also independent of the volitional control systems used to control waking visual perception. Lucid dreaming is a special type of "dreaming" that physiologically resembles waking and sleep meditative states. Lucid dreaming can conceptually delineate the special nature of volitional conscious control. Dream visual processing clearly utilizes the imagery operative cascade (see Figure 1.1). This operative protocol is a descriptive paradigm of cognitive processes that control the development of dream imagery. The dreamer is present as the point of view. Most dreaming imagery is viewed from the perspective of the dreamer, occurring to the left, to the right, in front of, or behind the dreamer's point of perspective in the dream. Most dreams are egocentric in nature, based on the point of view of the dreamer. This egocentricity includes a general sense of one's body as a bounded object located within a space containing other objects.

Dream imagery is most often in motion. The same CNS visual systems that recognize sensory patterns of object motion are also active during dreaming. The field of the image or a portion of the image can be altered under the direction of the dreamer. Image controls active during dreaming include the tendency to follow through to the completion of motions, the ability to scan and zoom in on an image, the ability to variably focus on the dreamscape, as well as the integration and manipulation of objects in relation to one another. The dreamer consciously controls this system with the use of focused attention, and the selection and integration of associative memories that change and alter the course of dream narrative. Dream images are emotionally and cognitively penetrated. Dreams often include an individual's beliefs, attributions, or expectations regarding a particular mental image. These images can be combined and altered to produce creatively unexpected emergent forms.

None of these consciously controlled processes is typically under the volitional control of the dreamer. From a film-based perspective, it is as though the dream were being directed by a second unit director creating a series of loosely connected vignettes. Without the waking control of a volitional and controlling executive director, the dream may or may not fit into the individual's overall life plot. The dream clearly does not include levels of executive production needed to integrate the dream into social, economic, and behavioral mores.

A form of the subconscious visual buffer is also present in dreaming. Components of dream imagery can quickly "grab" our mental attention, reminiscent of the way it is subconsciously grabbed by emotionally competent stimuli during perception (40). In dreaming, just as in waking, this visual buffer is particularly susceptible to emotional memories capable of triggering our attention. As discussed in Chapter 7, emotional processing functions during dreaming, with nightmares a likely symptom of dysfunction in this system. Central (contextual) images often provide a picture context for the emotions of the dreamer (41). Visual stimuli of sufficient emotional import have the ability to trigger associated memories, co-opting the plot line of the dream. Emotions are most often tied to images, and these "salient" dream images are those most likely to be remembered in waking (42).

THE VISUAL DREAM STAGE

The sensor (eye) and volitional conscious control have abbreviated and limited function during dreaming. During visual dreaming, it is the representational imagery system that achieves cognitive prominence. Dreams are a series of visual scenes, each comprised of representational images, tied together into narrative forms that utilize the internal formats (I-languages and grammar) of both thought and vision. The representational images as well as their language and grammar are often the same for different individuals. Some of these images are mundane, object representations such as "chair" or "horse." Others may be philosophically and spiritually profound: archetypical tidal waves, circles that could be the centers of the soul, or burning fires that could be interpreted as a representation of an individual's god or devil.

To this point, there is no concrete neuroanatomical evidence that the signs and symbols from dreams form into topographic images that are apparent in the visual cortex. Scanning systems do show increased activity in visual areas of the cortex during the sleep states of REMS and sleep onset in which dreaming is most likely to be occurring (43). But none of the current scanning systems has the capacity for detail to show representational patterns of neuron firing in the cortex. It seems likely, however, that dream images occur in association with topographic images that are projected neurologically in the CNS during dreaming. With future improvements in scanning detail, the study that concretely demonstrates this neuroanatomical correlation is likely to be one that studies video-gamers working with repetitive images and forms. The video game Tetris requires the repetitive operative control of geometric forms, a process that is somewhat analogous to that in which the monkey was trained to stare at a geometric target. When you wake Tetris players after sleep onset, for at least the next two hours they report that their dreams include Tetris-based geometric forms (44). Based on our current understanding of topographic brain mapping, it seems reasonable to speculate that such forms may actually be materializing and then dematerializing repetitively as neural patterns forming in the visual cortex of their brains.

This visual space of dreaming varies with the stage of sleep. Sleep-onset dreams are more intensely visual than other dream experiences. They can include components of exceptional vivacity and even hallucinatory reality. Although, as dreams, these experiences are non-perceptual, the visual system is functioning in sleep onset at a very different level from in the dreams of other sleep stages. It is as if the digital non-topographic systems of visual processing have yet to fully shut down. In sleep-onset dreams vivid visual reflections of the exterior world are still present. While less complex and varied than the visual experiences of waking perception, sleep-onset dreams are visually more vivid than dreams from other stages of sleep.

The subconscious visual buffer system is especially sensitive to emotional stimuli. Emotionally dominated dreaming is typical of nightmares, sleep paralysis (REMS), panic attacks (stage 2), and the parasomnias of deep sleep. Dysfunctions in the system of visual buffering could account for the primary symptoms of post-traumatic stress disorder (PTSD). In sleep, these are repetitive, emotionally disturbing nightmares and, in waking, a hyperresponse to intrusive stimuli. Disturbing nightmares are exquisitely sensitive to treatment using the cognitive restructuring therapies of eye movement desensitization and reprocessing (EMDR), imagery, and exposure (45). EMDR is an approach that consciously parodies the eye movements of REMS. EMDR functions subconsciously as a therapeutic approach known to improve waking function for a wide variety of psychiatric illnesses (46). Imagery therapy is most effective when used to treat trauma-associated nightmares in children. In children a visual approach to imagery therapy is used in which nightmare "monsters" are altered with colored pencils or crayons by the child (47). Both of these approaches used to treat disturbing nightmares are likely to use the visual buffer system to access the emotional processing system for cognitive restructuring. The visual buffer system apparently functions as an emotional cognitive interface. It is one of the few CNS cognitive processing systems that demonstrates such a capacity. We can use the visual buffer to cognitively affect and alter the emotional processing taking place during dreaming and nightmares.

Dream content is characterized by its continuity with our waking life, particularly the content characteristic of REMS and stage 2 sleep. While some content is propositional, there are clearly episodes of waking visual sensory perception and experience that find their way into these dreams. It has been suggested that dreams, particularly those of REMS, may function in learning and memory (see Chapter 3). There is some evidence for sleep having this function, but there are almost no studies suggesting that dreaming is a cognitive reflection of the incorporation of waking sensory experience into our memory during sleep (48). Experiences of continuity with waking occur less often in the dreams of sleep onset. They are not characteristic of deep-sleep dreaming. The dreams of stage 2 and REMS maintain closer ties with both the thought and visual contents of our waking experience than do dreams from the other stages.

While there are visual differences in dreams based on their form of consciousness, the visual aspects of most dreams have consistent characteristics whether the dream occurs in REMS, stage 2, sleep onset, or deep sleep. Visual dream imagery occurs from the perspective and point of view of the dreamer. Images are representational, based primarily on a shared visual I-language of intrinsic memory. The information in these images is limited compared with images based on external sensory perception. The visual imagery is processed in a large and extensively connected neuroanatomical staging area of the brain that develops these images internally in a representational fashion. These are cognitively penetrated images tied to emotions, and include beliefs, attributions, or expectations that can be creatively combined and altered to produce unexpected emergent forms. These dream images are ephemeral and change quickly on a millisecond time flux.

This neurovisual dream stage does not exist in isolation. Visual dreaming is complex and ever changing. The daily perceptions of vision are cognitively integrated and processed on a dream stage that assumes varied form and neural connectivity based on the stage of sleep. It fluxes and changes with background electrical and magnetic fields. On that stage, memories are incorporated, emotions processed, and narratives developed. Within the dream these elements re-form into new and alternative paradigms.

As visual beings, we are often best able to approach such a complex process with an image. Look back to the cover art of his book: the monotype and the calligraphy. These are artists' attempts at visualizing a dream.

Notes

1. Shakespeare, W. (1606/1975) Macbeth, *The Complete Works of William Shakespeare*. New York: Gramercy Books, p. 177.
2. Hartmann, E. (2011) *The Nature and Functions of Dreaming*. Oxford: Oxford University Press, p. 12.
3. Jung, C. (1948/1974) On the nature of dreams, in *Dreams*, trans. R. Hull. Princeton, NJ: Princeton University Press, p. 68.
4. Roediger, H., Weldon, M. and Challis, B. (1989) Explaining dissociations between implicit and explicit measures of retention: A processing account, in *Interference and Cognition*, ed. F. Dempster and C. Brainerd. New York: Academic Press, pp. 29–59.
5. Mausfeld, R. (2003) Conjoint representations and the mental capacity for multiple simultaneous perspectives, in *Looking Into Pictures: An Interdisciplinary Approach to Pictorial Space*. Cambridge, MA: MIT Press, pp. 32–33.
6. Kosslyn, S., Thompson, W. and Ganis, G. (2006) *The Case for Mental Imagery*. Oxford: Oxford University Press, pp. 14, 135.
7. States, B. (1997) *Seeing in the Dark: Reflections on Dreams and Dreaming*. New Haven, CT: Yale University Press, p. 97.
8. Scarry, E. (1995) On vivacity: The difference between daydreaming and imagining-under-authorial-instruction. *Representations* 52: 1–26.
9. Sartre, J.-P. (1940/1991) The imaginary life, in *The Psychology of Imagination*. New York: Citadel Press.
10. Scarry, E. (1999) *Dreaming by the Book*. Princeton, NJ: Princeton University Press, p. 3.
11. Kosslyn, S., et al. (2006) *The Case for Mental Imagery*, p. 57.
12. Finke, R. and Shepard, R. (1986) Visual functions of mental imagery, in *Handbook of Perception and Human Performance*, ed. K. Boff, L. Kaufman and J. Thomas. New York: Wiley-Interscience, p. 37; Thompson, W. and Kosslyn, S. (2000) Neural systems activated during visual mental imagery: A review and meta-analysis, in *Brain Mapping: The Systems*, ed. A. Toga and J. Mazziotta. San Diego, CA: Academic Press, pp. 535–560.
13. Kosslyn, S., et al. (2006) *The Case for Mental Imagery*, p. 100.
14. Zeki, S. and Bartels, A. (1999) Toward a theory of visual consciousness. *Consciousness and Cognition* 8: 225–259.
15. Gregory, R. (1997) *Eye and Brain – The Psychology of Seeing*. Princeton, NJ: Princeton University Press, pp. 76–77.

16. McCarthy, R. (1993) Introduction: What and where, in *Spatial Representation – Problems in Philosophy and Psychology*, ed. N. Eilan, R. McCarthy and B. Brewer. Oxford: Oxford University Press., pp. 319–324.
17. Gregory, R. (1997) *Eye and Brain*, pp. 76–77.
18. Fox, P. T., Mintum, M. A., Raichle, M. E., Miezin, F. M., Allman, J. M. and Van Essen, D. C. (1986) Mapping human visual cortex with positron emission tomography. *Nature* 323: 806–809; Felleman, D. and Van Essen, D. (1991) Distributed hierarchical processing in primate cerebral cortex. *Cerebral Cortex* 1: 1–47. Sereno, M., Pitzalis, S. and Martinez, A. (2001) Mapping of contralateral space in retinotropic coordinates by a parietal cortical area in humans. *Science* 294: 1350–1354.
19. Tootell, R., Silverman, M., Switkes, E. and De Valois, R. (1982) Deoxyglucose analysis of retinotopic organization in primate striate cortex. *Science* 218, 902–904.
20. Kosslyn, S. (1994) *Image and Brain – The Resolution of the Imagery Debate*. A Bradford Book. Cambridge, MA: MIT Press.
21. Finke, R. (1990) *Creative Imagery: Discoveries and Inventions in Visualization*. Hillsdale, NJ: Erlbaum.
22. Sachs-Hombach, K. (2003) Resemblance reconceived, in *Looking into Pictures: an Interdisciplinary Approach to Pictorial Space*, ed. H. Hetch, R. Swartz and M. Atherton. Cambridge, MA: MIT Press., pp. 167–178.
23. Gregory, R. (1997) *Eye and Brain*, p. 116.
24. Chanan, M. (1996) *The Dream That Kicks: The Prehistory and Early Years of Cinema in Britain*. London: Routledge.
25. Klatsky, R. and Lederman, S. (1993) Spatial and non-spatial avenues to object recognition by the human haptic system, in *Philosophy and Psychology*, ed. N. Eilan, R. McCarthy and B. Brewer. Oxford: Oxford University Press, pp. 191–205.
26. Rodowick, D. N. (1997) *Giles Deleuze's Time Machine*. Durham, NC: Duke University Press.
27. Vuilleumier, P. and Swartz, S. (2001) Beware and be aware: Capture of spatial attention by fear-related stimuli in neglect. *NeuroReport* 12: 1119–1122.
28. Eilan N., McCarthy R., Brewer B. (ed.) (1993) *Spatial Representation: Problems in Philosophy and Psychology*. Oxford: Oxford University Press.
29. Kosslyn, S. (1994) *Image and Brain*.
30. Zeki, S. (1999) *Inner Vision: an Exploration of Art and the Brain*. Oxford: Oxford University Press, pp. 66–67.
31. Mausfield, R. (2003) Conjoint representations, in *Looking into Pictures: An Interdisciplinary Approach to Pictorial Space*, ed. H. Hetch, R. Swartz and M. Atherton. Cambridge, MA: MIT Press, pp. 17–60.
32. Hopkins, R. (2003) Perspective, convention, and compromise, in *Looking into Pictures: An Interdisciplinary Approach to Pictorial Space*, ed. H. Hetch, R. Swartz and M. Atherton. Cambridge, MA: MIT Press, pp. 145–166.
33. Collier, J. and Collier, M. (1912/1986) *Visual Anthropology: Photography as a Research Method*. Albuquerque, NM: University of New Mexico Press.
34. Hopkins, R. (2003) Perspective, convention and compromise.
35. Treisman, A. (2000) The binding problem, in *Findings and Current Opinion in Cognitive Neuroscience*, ed. L. Squire and S. Kosslyn. Cambridge, MA: MIT Press, pp. 31–38.
36. Wollheim, R. (2003) In defense of seeing in, in *Looking into Pictures: An Interdisciplinary Approach to Pictorial Space*, ed. H. Hecht, R. Schwartz and M. Atherton. Cambridge, MA: MIT Press, p. 5.
37. Marks, D. (1990) On the relationship between imagery, body and mind, in *Imagery – Current Developments*, ed. P. Hampson, D. Marks and J. Richardson. New York: Routledge, pp. 1–36. Pagel, J. F. (2008) *The Limits of Dream: A Scientific Exploration of the Mind/Brain Interface*. Oxford: Academic Press, pp. 157–158.

38. Dement, W. and Vaughan, C. (2000) *The Promise of Sleep*. New York: Dell.
39. Fitt, A. and Gonzalez, G. (2006) Fluid mechanics of the human eye: Aqueous humour flow in the anterior chamber. *Bulletin of Mathematical Biology* 68: 53–71.
40. Kosslyn, S. (1994).
41. Hartmann, E. (2011) *The Nature and Functions of Dreaming*, p. 12.
42. Kuiken, D. and Sikora, S. (1993) The impact of dreams on waking thoughts and feelings, in *The Functions of Dreaming*, ed. A. Moffitt, M. Kramer and R. Hoffman. Albany, NY: SUNY Press., pp. 419–476.
43. Semba, K. (2011) Preoptic and basal forebrain modulation of REM sleep, *in REM Sleep: Regulation and Function*, ed. B. Mallick, S. Pandi-Perumal, R. McCarley and A. Morrison. Cambridge: Cambridge University Press, pp. 99–109.
44. Rittenhouse, C., Stickgold, R. and Hobson, J. (1994) Constraints on the transformation of characters, objects, and settings in dream reports. *Consciousness and Cognition* 3: 100–113.
45. Pagel, J. (2010) Drugs, dreams and nightmares, in *Dreaming and Nightmares*, ed. J. F. Pagel, *Sleep Medicine Clinics* 5(2). Philadelphia, PA: Saunders/Elsevier, pp. 277–288.
46. Shapiro, F. and Forrest, M. (1997) *EMDR: Eye Movement Desensitization and Reprocessing*. New York: Basic Books.
47. Siegel, A. and Buckley, K. (1998) *Dream Catching: Every Parent's Guide to Exploring and Understanding Children's Dreams and Nightmares*. New York: Three Rivers.
48. Smith, C. (2010) Sleep states, memory processing, and dreams, in *Dreaming and Nightmares*, ed. J. F. Pagel, *Sleep Medicine Clinics* 5(2). Philadelphia, PA: Saunders/Elsevier, pp. 217–228.

Creating Artificial Dreams

... why should I not grant to dreams what I occasionally refuse reality, that is, this value of certainty in itself which, in its own time, is not open to my repudiation? **(Breton, 1924) (1)**

Many artists create simulacrums of their dreams. This is not a new approach. As noted repeatedly in the first chapters of this book, there is good if indirect evidence that dreams had a role in the creation of the most ancient of artistic masterpieces – the cave paintings of south-west Europe. Historically, with the development of any new technology that can be used to create art, the first documented attempts in that medium are often attempts to recreate dreams. Recorded dreams were discovered among the first decipherable attempts at writing (see Chapter 2). Initial attempts at film-making included dream-like "visages." At the beginning of the last century, the Lumière Brothers, G. A. Smith, and the Edison Labs all created among their first films those titled as *Dreams*, foggy and jerky, black and white moving pictures that somehow resemble dream imagery more than waking perception (2). From the first, film as a visual medium has often been seen as an attempt at creating apparent if artificial dreams.

Dream Science.
DOI: http://dx.doi.org/10.1016/B978-0-12-404648-1.00010-7

Typically, in their study of dreams, scientists have followed the approaches used by artists. Ancient shamans developed their role by conducting ceremonies that included the images painted in the caves. Much later, physicists studied the perspective system and the color wheel adopted by Renaissance artists in their attempts to understand the science of optics. The mechanisms of photographic fixation led to photochemistry and eventually to photovoltaic sensors. Neuroscientists study the techniques of film in their attempts to understand the cognitive processing utilized for sensory vision as well as for mental imagery. Psychoanalysts and psychiatrists used the process of film-making as a model for developing theories that might explain the basic structure of mental dynamics.

FILM-MAKING: THE MENTAL APPARATUS

The visual process of film-making developed on a trial-and-error basis. The applied techniques that worked led to the science of optics and to an understanding of cognitive visual processing, as well as contributing to the advent of psychoanalysis. Freud set down his theories during the same era that the techniques of cinema were being developed. He realized that in cinema the spectator was in a psychodynamic relationship with the camera and screen. The theoretical basis of psychoanalysis is based in part on Freud's "apparatus" theory, also called the theory of "psychical locality" (3):

> I propose simply to follow the suggestion that we should picture the instrument which carries out our mental functions as resembling a microscope or photographic apparatus, or something of the kind. On that basis, psychical locality will correspond to a point outside the apparatus at which one of the preliminary stages of an image comes into being. In the microscope and telescope, as we know, these occur in part at ideal points, regions in which no tangible components of the apparatus is situated …. Accordingly, we will picture the mental apparatus as a compound instrument, to the components of which we will give the name "agencies," or (for the sake of greater clarity) "systems." It is to be anticipated, in the next place, that these systems may perhaps stand in a regular spatial relation to one another, in the same kind of way in which the various systems of lens in a telescope are arranged one behind the other. (*Freud, 1914*) (*4*)

The individual is not fully present or in volitional control during dreaming. The viewer is an onlooker to his own dream, enveloped by the dream just as a child is enveloped by his world. This is a world view that is often filled with magic. The subject does not see where a dream is leading, yet he or she follows (5). The experience of viewing cinema shares this characteristic of non-volitional envelopment. Through cinema, the film-maker apparently taps the ability of dream imagery to assess the unconscious mind. Apparatus theory describes what some have called a "cognitive machine," a map for the psychoanalytic formulations of character, genre, and illusion that can be used by both the psychiatrist and the film-maker (6).

Jacques Lacan expanded on Freud's apparatus theory to develop his own analytic model of the psyche. Lacan based his psychoanalytic construct of the *Other* on the ideal of our self that we see reflected in a mirror. In viewing our projection in the mirror we are looking at a dramatically ambivalent understanding of ourselves, an ideal self-image that is always outside us – the Other. Lacan proposed that the power of cinema came from tapping this interior psychic confusion between self and this projected Other (7). It is based on our ability to contemplate our mirror-image self that we are able to identify with the visual images that we encounter in the cinema. We can identify alternatively with exhibitionist and the voyeur, the master and the slave, the victim and the victimizer. As spectators, we are in a dynamic where the apparent mirror of the film leads us to believe that we are present and involved with the images on the screen (8). Through this process of identification we are both enmeshed in and displaced from our identification with the Other, a process that demands sameness and similarity while disallowing difference (9). This perspective of the mind as a mental apparatus that approximates the cinematic camera has had considerable resonance for psychology, film theory, and popular culture. Viewer identification extends beyond the cinema to social and economic systems, in which films and actors are often used in entertainment, product, and political marketing. The constructs of apparatus theory, mirroring, and identification are psychoanalytic constructs that can be used to explain the power of film in our personal, social, and political lives.

Film utilizes at least some of the same cognitive processing systems that are part of dreaming. Cinematic language often provides a better handle for describing the phenomenology of dreams than do the sciences of linguistics, optics, electrophysiology, neurology, or the other philosophies that have been used to describe dreaming. Terminology adopted from film includes eye-as-camera, projection, point of view, crossing the line, the subordination of time to motion, and flashback. Cinematographic techniques imitate the operative process that we use to visually organize imagery and perceptual experience (see Table 1.1 in Chapter 1). Like dreaming, films have the capacity to create an almost complete cognitive experience, fully outside the viewer's control. They can affect our memory systems, our visual imagery, and our emotions. The specific images presented on film can be hard to remember in the same way that dreams are hard to recall. The visual storyline of a film is composed of associated images that are able to interact with the personal memories and emotions of the viewer. You find yourself remembering moments of a film, trying to find your way back to the mood and associations associated with a particular moment, tapping into emotionally competent stimuli that led you down personal story-paths far different from those planned by the film-maker. Films can provoke both nightmares and ecstasy – the extreme emotions of the dreamer faced with his own creation (10). As in dreams, certain moments of films viewed decades ago can be as vivid as moments of childhood, treasured moments that serve as a baseline comparison to waking experience (11). In films as in dreams, everything takes place in the present tense, an always initial experience, no matter how often you have been to the same screening or had the same dream. The film-maker can play games with the viewer, presenting to and incorporating the audience in an alternative reality. And like dreaming, a film is sometimes better understood later, after being subjected to an independently conscious process of translation and interpretation. The complex intellectual methodology of Freudian and post-Freudian film interpretation and theory has become the current "orthodoxy" of most university cinema study programs.

In attempting to understand the techniques involved in creating an artificial dream, it is useful to examine the way in which a film-maker creates a cinematic dream. Early cinema was constructed around a stationary camera that was set up to be an audience for the actor. Early films were set-pieces of stage performance. Clips from that era have the energy of on-stage burlesque actors who filled the camera with a wild uninterrupted flow of gestures. The redirection of the camera from body shots to the close-up gaze of a character created a revolution in early cinema. Actors developed techniques such as immobility and repression of gesture expressly for the new medium. Film-makers began to use the close-up as a technique for pulling the spectator into emotional contact with a character. The carnivalesque, joyful, and dirty elements of early cinema were replaced by an emphasis on interior emotions. According to some, it was this close-up focus on the gaze, rather than minions of the devil, that led film-makers astray and brought desire, perversion, and obsession to the screen (12).

Film-makers create suspense in order to keep the viewer's attention. They use editing and cuts of parallel actions: interspersing the cut-in image of a knife approaching a bared throat against one of a car racing along a road in a cloud of dust, to bring the audience to a level of anxiety as to whether an anticipated rescue might occur. Events that occurred over only a very short time can be expanded with editing and slow-motion filming to develop a crescendo of suspense. The shower scene in Hitchcock's *Psycho* is an excellent example (13). Film-makers have other techniques that can be used to induce suspense. One approach is to use a close-up in a scene in which the viewer would expect an establishing shot. Confounding viewer expectations can add suspense. Orchestration and ambient sound can be used to affect mood as well as to confound expectations.

Narrative flow can be altered and expanded with editing. Master shots brought together in parallel and then into focus can be used to bring together disparate tracks of the storyline. Detail shots can internalize psychological themes and structures for the viewer. And then, of course, there is the ubiquitous flashback, an interjected scene that takes the narrative back in time from the current point, and the perennial topic of fascination for both viewers and analysts. D. W. Griffith first used flashbacks in *Intolerance* (1916), presenting a single shot of a mother rocking a cradle, repeated many times between scenes to mark the passing of generations (14). Formalism, structuralism, semiotics, theories of ideology, philosophies of memory and consciousness, as well as psychoanalysis, have been used in the attempt to understand the power of the viewer's response to the flashback. Carl Jung adopted the concept of flashbacks into his psychodynamic theories using the term *internal analepsis* to denote a flashback to an earlier point in the narrative, and *external analepsis* to refer to a flashback to a time before the narrative or psychotherapy session had started (15). More than any other concept, the flashback highlights the remarkable interplay between film dynamic and psychoanalytic concept.

We have been trained as viewers to expect that a film will include accepted film-making conventions. The experience of film viewing requires audience familiarity with the technical procedures utilized in narration, genre, and projected illusions. Famously, the audience watching one of the first moving pictures of an oncoming train is reputed to have dodged out of the way. When we watch a film we now maintain the unstated awareness that we are watching a perceptual illusion that we pretend to be real. The spatial parameters of visual shots are embedded within the causal logic of a story. We maintain a continuous line of narrative action from our perspective as trained viewers. Transitions between shots are disguised by the film-maker and ignored by the viewer so that image and sound are experienced as a continuous present moving forward in time. While the director of a film may attempt to subvert our expectations of narrative, genre, pace, time, and character, it can be difficult to subvert an audience's expected illusion. Developed conventions, once learned by the viewer, can neutralize such attempts, with the interplay between film-maker and audience becoming a dance of expectations that are only sometimes fulfilled. In a film such as *Memento*, narrative structure is actually built on the audience expectation that it is based on a progressively developing plot. For the viewer, the plot persists even after it is subverted by the film-maker, who creates a character without short-term memory who must relearn his film reality every time that he wakes from sleep (16).

It was apparatus theory that first suggested that remembered dreams and the unconscious mind operate within the film-making dynamic.

Post-Freudians proposed that film-makers affect us emotionally and alter our perceptual view of the exterior universe by accessing the same mental systems that we utilize in our dreams. Film-makers developed complex methods of storytelling based on this perspective, intentionally attempting to connect that process with the anxieties of the audience. In his work, Lacan analyzed the imaginary and symbolic process of the film viewer. He called this process "suture," a medically based metaphor implying that through techniques such as editing, a film could be the "sewn" shut so as to include the viewer (17). The sutured viewer shifts from one character or scene to another following the eyelines (gaze) of the characters, accepting what is seen on film from multiple perspectives and directions as natural. In the classic example of shot/reverse shot editing, two characters are viewed alternatively over the other's shoulder. We do not ask "Who is watching?" because each shot answers the question of the previous shot (18). Dayan calls this the "tutor-code" of classical cinema: "Unable to see the workings of the code, the spectator is at its mercy. His Imaginary is sealed into the film" (19). We have become inured to such cinematographic conventions. When watching the rough cut of a film in which this code is broken and the sight line between characters slips, the viewer can become physically nauseated.

The psychodynamic characteristics of film are perhaps the best evidence that psychoanalytic theory has some basis in fact. Cinematic language is often used to describe the phenomenology of dreams. Operative approximations of aural and visual imagery are artificially reproduced during the making of a film. The film-maker recreates the cognitive operating systems of a dream. Today, cinematographic images created within the context of film narrative and sound are the closest that we can come to artificial dreams. Neither dreams nor film is under the conscious control of the viewer. Both are coded representations of an external reality. The two processes (dreams and film) are clearly interactive. Images from the films that we watch are often incorporated into our dreaming based on continuity with our lives, personal resonance, and the strength of the emotional stimuli (20). Dream scenes are often included as part of the storyline in films. They provide insight into the inner drives and psychological makeup of a character, and bring the viewer into the experience of the dream. The film-maker projects his or her dream on screen using this "bidirectional" characteristic of dreaming, sharing a filmatic dream in which an interior world is projected on to the screen in the attempt to affect external society and culture (21).

This film-based dream dynamic has a resonance and logic that seem to be real. But there are obvious differences between dreams and film. From the perspective of volitional control, dreamers do not know that they are dreaming, while spectators know that they are at the cinema. In film, both sound and presented visuals are directed and externally controlled.

Cinema includes both the transcendental subject and the object to be perceived by the viewing subject, a perspective that continues to exist with or without an actual viewer in the audience (22). The projected dream of cinema is an intentionally accepted delusion experienced when awake, with the viewer enmeshed into a film "reality" in an imaginary present tense (23). Cinema is a fully an audiovisual experience and includes none of the other perceptual sensations that are sometimes incorporated into dreaming such as smell, pain, or body sensation. Cinematic images are less personal and individual, and more social and ideological.

The concepts of film-making and those of psychoanalysis developed in concert, during the same era, and based on many of the same assumptions. This potentially accounts for the conceptual similarities between the fields. But a dream shares much with its projected counterpart. Films are made up of propositional images, outlines rather than fully formed objects that are very similar to the representational images projected on to the cells of the brain. These images in our brain change even more quickly than the present stills of a moving picture, appearing and disappearing at the speed of thought on a millisecond to millisecond basis. In rapid eye movement sleep (REMS), visual rapid eye movements have the capacity to mark chronological time, allowing for a narratively controlled chronological timeline. During both film and dreaming the visual buffer is active, preferentially processing emotions. The visual buffer can function as a varyingly conscious cognitive interface that can affect emotional processing during both wakefulness and sleep. Both films and dreams access neuroanatomically specific memory stores. The films that we watch lead us down a Purkinje tree of associated memories, just as in dreaming. Film, as an artificial construct, approximates many of the same biological systems that we utilize cognitively in dreams.

ARTIFICIAL CONSCIOUSNESS

The functional capabilities of artificial intelligence (AI) systems extend far beyond human capacities. Perceptually, sensors utilize input from well outside the human biological range to access visual, auditory, positional, and tactile data. The reasoning capacities of AI systems now extend far beyond the already remarkable computational abilities of digital systems. Probability and fuzzy reasoning analytic approaches requiring complex parallel and quantum processing computer systems can more closely approximate biological system processing, and allow these systems to deal with uncertainty in the information and processing of data stored in large knowledge bases (24). Intelligence, based on a human definition, consists of specifying a goal, assessing the current situation to see how it differs from that goal, and applying a set of operations that reduce the differences (25). AI systems can now be considered as intelligent. That intelligence is independent of their roles in facilitating human functioning (26). The Mars Landers highlight current AI system capabilities. They have traveled to and explored biologically toxic complex environments, generating their own energy while traveling, handling navigational tasks, and completing complex astronomical, climatic, and geological experiments, and then broadcasting that data back to human handlers on another planet. The complexity of the presented data, even after computerized attempts at analysis, is often beyond the monitoring humans' ability to rationally comprehend. Have the Landers discovered the presence or absence of life on Mars? Perhaps the Landers know.

Yet there is no clear evidence that these exceedingly complex and intelligent systems have achieved independent volitional consciousness. Some AI researchers have proposed that when a system becomes sufficiently complex and intelligent, self-awareness and other aspects of mind can be expected naturally to follow (27). These capabilities would not be programmed but would develop based on the increasingly complex interactions between programmed subsystems and net overload (28). Our current system of computerized phones and personal computers interconnected by transmission-line and broadcast interface achieves an extreme level of computational and processing complexity. Based on the complexity leading to consciousness hypothesis, this system could very well be conscious. However, despite the occasional unexpected power outage, and the unexplained viruses that corrupt our systems, there is little evidence that this extremely complex system has developed an independent volitional consciousness.

This has led some to approach this issue by redefining consciousness, an approach with a long and contentious history, as noted in Chapter 4. For some modern neuroscientists, with Christof Koch a primary spokesperson, the marker for consciousness is the capacity for independent decision making. This is a capability within the capacity of a photovoltaic switch that can respond to a change in ambient lighting by turning a system on and off (29). Faced with the amazing functional capacity of modern AI systems, some philosophers now define aspects of mind as those that we cannot scientifically explain. Aspects of mind are those aspects of cognitive functioning that we cannot artificially create (30).

Dreams, the essence of subjectivity, could mark the development of consciousness by AI systems, evidence for the development of an independent subjective component of mental process within the hardwired systems. Beyond being the marker for a leap to consciousness, dreams and dream-like phenomena have potential processing functions that could be useful for AI. Dreams could be utilized, as in biological systems, to expand the creative capabilities of AI – providing unexpected and alternative answers to difficult questions (see Chapter 1). This capacity for providing unexpected answers to exceedingly complex questions is, however, already within the capability of current AI systems. Weather/climate forecasting is an example in which a series of mathematical models is constructed around extended sets of dynamic equations known to govern atmospheric motions and turbulent diffusion, as well as the data analysis of radiation, moisture content, heat, vegetation, surface water, terrain, and convection. These equations, impossible to solve through analytical methods, require the complex computational and processing integration of powerful supercomputers. The accuracy of predictions varies with the density and quality of data, as well as any deficiencies and limitations inherent in the numerical models (31). The outcome derived from such a complex AI computing process is sometimes unexpected and often difficult to explain.

Such a complex developed analysis that provides unexpected and alternative answers to questions has some of the characteristics of a dream. Attained results that diverge from expectations can almost be considered "creative." Such an analysis, like dreaming, integrates extensive sensory data. Both require the associative interactions of many processing subsystems. The complex results of these analyses are often incomprehensible except when presented as a visual display changing through time, somewhat like the remembered dream. Result analysis, like dream interpretation, is often a metaphoric and allegoric process affected by the training and belief systems of the researchers. Such "AI dreams" share at least these characteristics with their biological counterparts.

THE INTERFACE THAT ABROGATES CONSCIOUSNESS

Interfaces facilitate and/or extend the interaction between our brain and the external environment. Beyond the punch card and keyboard, the biological systems of eye and papillary motion, tactile and proprioceptive sensation, auditory command, and electrophysical receptors have all proven amenable to interfacing with AI systems. The primary role of an interface is to attempt to imitate or improve upon biological sensory perceptual capabilities. Interface capacity can be extended by incorporating cognitive processing systems similar to those utilized by the central nervous system (CNS). For example, systems with auditory recognition capacity also integrate the ability to phonetically understand, translate, semantically represent, and classify the representations of sound and human speech. These processes must occur well before the ambient sounds can be conceptually manipulated. Computerized "war games" exemplify interface interactions. Weapons are virtually aimed and soldiers are trained on war game computer systems that utilize a complex visual, vocal, tactile, and auditory interactive interface. In the aiming of weapon systems from drones and mobile robots, we have reached the point where the experience of warfare includes virtual and real death on screen. At its most basic level, consciousness is determined as life versus death – the presence or persistent absence of response to the external environment. The military application of such war "gaming" systems achieves a final concrete conflation of computerized representation with terminal reality. It is not surprising that we are prone to confuse the concept of mind with the brain process of sensory integration, a capacity within the capability of computer systems. For newer systems, the controller is no longer needed to exert the control leading to weapon-release. That decision can be preprogrammed based on sensory input and the computerized analysis of the acquired target. The decision of life or death, the decision to abrogate a specific consciousness, is made by the AI system without prompting or the requirement for a human interface.

Dreams function as the cognitive interface between waking and sleeping forms of consciousness. Our dreams present to our waking state information about the mental activity occurring within the forms of consciousness present during sleep. This information would otherwise be unavailable to our waking volitional consciousness and thought. During sleep, non-conscious systems including electrophysiological, neurochemical synaptic, neuroendocrine, and neural membrane messaging systems are functioning in the CNS. From within the neuron, the information incorporated into the DNA genome controls cellular processing, probably affecting shared systems of language and semiotic processing such as narrative grammar and syntax (see Chapter 8). This information is also likely to include the intrinsic memories and shared images that are incorporated conceptually into primitives and archetypes (see Chapter 9). During waking, specific information is available from these systems primarily as remembered dreams from the sleeping forms of consciousness.

Dreaming functions as the interface between these varyingly conscious brain states. This has led some to postulate that dreaming could function extracorporally (32). This is a commonly accepted belief. Many modern religions share the belief that spiritual messages and insights can be received from outside oneself during dreaming (33). A majority of adults indicate that they have personally experienced dream content-based *déjà vu* and the foretelling of future events (34). There is even some research suggesting that dream images can be broadcast and transmitted from one person to another (35). While it has been difficult to demonstrate scientifically that such events actually occur, there is obviously much about dreaming and the functioning of the human mind that scientists have yet to understand.

In the 1950s, it was proposed that sleep-learning (hypnopedia) might be possible, and several studies were published that seemed to demonstrate learning occurring during sleep. However, researchers discovered that sleep-learning could only be documented in those subjects in which alpha-wave activity was present on electroencephalography (EEG). Since alpha is the marker for drowsy wakefulness, it was obvious that for these subjects learning occurred only during behaviorally apparent sleep that was actually an eyes-closed awake state (36). A more recent study indicates that some types of classical stimulus–response conditioning can take place during sleep. Subjects can be taught during sleep to develop new associations between tones and odor. This acquired behavior persists into wakefulness without conscious awareness of the learning process (37). This is far less than the in-depth hypnopedia postulated by so many science-fiction authors, but it does indicate that during at least some stages of sleep, auditory and olfactory input can be processed and learned.

Humans are visual creatures. When learning and representing information, we generally think about and comprehend complex patterns of information using a visual modality (38). The visual perceptual isolation of sleep is the primary limiting factor preventing dreaming from functioning as an external cognitive interface. Yet visual processing systems are active during dreaming. These systems include the emotionally sensitive visual buffer system and the important topographic imagery system. The subconscious buffer system functions during waking as a cognitive interface, and is one of the few waking interfaces that is sensitive to non-cognitive approaches such as eye movement desensitization and reprocessing (EMDR). The propositional visual system of topographic imagery functions in its most indomitable form during dreaming. As the neuroscientist Rodolfo Llinas has pointed out, wakefulness has its own limitations. It may actually be a kind of dreaming in which neural activity is constrained by sensory input (39). Viewing all consciousness as dreaming, it is a smaller step to the postulate that dreams could be extracorporally shared (40). Some scientists see this as but one of the potential future paths that could alter our species from one functioning as an independent collection of individuals to one sharing a transhuman collective consciousness (41).

DREAMTIME ELMO

This is an era of technology. A simulacrum of almost anything that is conceptually understood can be artificially created. In this book, the biological components of dreaming have been addressed in detail. A dream is constructed out of constituent elements, primarily memories, emotions, and visual imagery. Dreams are most often experienced visually, and the visual perceptual system is the best understood and best described of the CNS cognitive processing systems. Dreams have continuity with our lives based on day-residue memories of waking experience. The specific memories incorporated into dreams are neuroanatomically based, and built with the same memory processing systems as those used to form waking perceptual memories. Emotional expression is also brain based, controlled by distinct neuronal circuits that are integrated and connected with motor and perceptual systems. Feelings process and integrate these emotions with emotionally competent memory systems. Feelings and emotions alter and affect an extensive network of motor, sympathetic, endocrine, and parasympathetic expression that extends throughout the body. While complex, the dream-associated system of visual images, emotions, and memories is based on well-described neuroanatomy and neural processing. All of these technical paradigms of the dreaming brain: digital on–off neuron connections, memory storage, visual operative processing, interactive messaging systems, and emotional triggering/buffering systems operating at subliminal speeds, can or could be artificially constructed. We have already considered the possibility of Mary Shelley's *Frankenstein*. Beyond "Tickle Me," consider the possibility of "Dreamtime" Elmo.

The images, memories, and emotions of dreaming are formed into a narrative structure and presented on awakening as a story. This narrative structure is both easier and harder to artificially create than these other biological parameters of dreams. Dreams are presented as stories using the same structuring methods and principles that we use for structuring waking thought. Authors often include dreams in their texts and films. There is a long history of such work, and there are well-described techniques for both creating and using the dream-like narrative as a technique to pull the reader or viewer into the inner world of a character. These techniques can be used to create for the reader a construct that makes the described experience look, sound, or feel the way that it would if it were a dream. We almost all experience dreaming, and the presented world of the artificial dream reflects our own experience of the dream state. In our viewing or reading the constructed dream, we interject our own memories, emotions, and imagery into the experience. And once enmeshed, once sutured into the inner world of that scene and that character, our cognitive experience extends beyond the experience of the artificial dreamscape as a dream. We may experience something far different from what was designed or plotted by the writer. The writer can entice the reader or viewer toward the vicarious experience of mental images that resemble less a daydream and more the vivacity of actual experience (42). It is this possibility, coupled with the potential role for dreaming in creative problem solving, that has attracted the creators of AI systems to the area of dream science. For an AI system, dreaming could mark the development of a reflexive and independently volitional consciousness.

During both waking and sleep, the forms of consciousness are expressed from within the electrophysiological, neurochemical, and neuroanatomical framework that makes up the human CNS. Through the external interfaces of sensory perception and motor activity we integrate our entire spectrum of CNS processing with that of others, our society and culture, as well as with the external environment. The human CNS is, without a doubt, the most complex biological system that we have attempted to describe and understand. Its neuroanatomy can be described at different levels, each built to scale. There is the organ: the global interconnected brain that has multilevel connections to all areas of the body. The brain is parsed into components, areas of specific function that can be visually examined and delineated. On the microscopic level, the brain includes more than 100 billion neurons. Through neural processes, each has multiple neural net connections. At those synapses, each neuron can respond to multiple neurotransmitters and modulators. Each has access to the vascular system and is bathed in spinal fluid, existing and functioning in a complex potion of extracellular chemicals that affect cellular receptors, synaptic firing, metabolism, and electrolyte concentrations. Most neurons are supported by helper cells that include astrocytes and electrically responsive glial cells. A cell is a potentially independent organism that has developed to assume a structural and functional role within the multicellular being. On the electromicroscopic level, each cell is a small world of incredible complexity with metabolic, communication, and memory systems that have decision-making capabilities. An individual cell can become ill, procreate, and even die. In the nucleus of each cell is the repository of codes and shared information that exists within our DNA, the most complex of biological molecules, and one able to express an exquisite electrical and chemical sensitivity and response to its environment. The neuron functions within an environment of extracellular electrical and magnetic fields that change and flux on a millisecond basis. These fields are known to affect cellular equilibrium, intracellular proteins, metabolism, and the tendency of each neuron to fire. They may affect the expression of DNA. Our ability to consider, describe, and conceptually attempt to understand such a complex system is excellent evidence for the incredible integrative capacity of the human CNS.

As noted, our culture is technologically adept. While exceedingly complicated, much of this biological framework could be created artificially – at least in components. Consider the possibility of 100 billion computer chips, an AI correlate for each of the neurons comprising the human brain. Each could be constructed with similar basic programming, varyingly coded memories, images, and emotions. Each could be interconnected to the electrical and chemical environment through an interface of similar complexity.

Would such a system be conscious? The answer is definition dependent, and consciousness, as noted previously, is poorly defined. But from most perspectives the answer to this question is likely to be "yes." Biological systems of lesser complexity are clearly conscious. Why would a non-biological system of similar construction not be conscious? Current systems have volitional capabilities for decision making, as well as the capabilities for self-evaluation, self-programming, and self-repair. Based on most definitions of consciousness (see Table 4.1 in Chapter 4), AI systems have either already achieved or could conceptually achieve consciousness. That attained level of consciousness is not limited to the primary capacity of reactivity to the environment. AI can process and respond to perceptual input and make decisions in a logical manner. AI has demonstrated the capacity to abrogate consciousness. There are AI systems with the ability to achieve forms of tertiary consciousness such as cognitive feedback, self-recognition, and components of self-concept (43). There are, as discussed in this book, many forms of consciousness, and a better question might be: Can AI achieve an equivalent to human consciousness?

One way to phrase this question is to ask: Could such a system dream? As already noted in this chapter, there are AI equivalents to dreaming. Dreaming has the characteristics, based on definition (see Chapter 3), of existing on a sleep–wake continuum (axis 1), reported recall (axis 2), and forms of content (axis 3). Based on the first axis, by definition, AI systems would be required to develop the capacity for sleep. Sleep is behaviorally defined as a state of reversible perceptual isolation. That is easy enough to create for an AI system using something as simple as an on/off timer, but as noted in Chapter 6, sleep is anything but a cognitive state of null activity. Sleep is a state of perceptual isolation that, while otherwise functional, is an operational capacity easily within the capability of AI. Current systems like the Mars Landers are programmed to have periods of quiescence, during which sensors are turned off while other processing systems remain in operation. Data/content are reported from such periods of non-perceptual isolation (axis 2). Like most dreams, that content has continuity with sensory recording obtained during periods of full operational capacity (wake). These systems report data (content) during periods of perceptual isolation (axis 3). On this basic level, current AI systems can easily meet the sleep/wake, report, and content criteria required by the accepted operational definition of "dreaming" (see Box 8.2 in Chapter 8).

This is scientific and empirical evidence indicating that AI now has the capacity to meet definition criteria for both consciousness and dreaming. Aspects of mind are within the capacity of current AI systems. Constructs of emotion, feeling, creative insight, and expression can be artificially created. There are artificial equivalents for primary consciousness, secondary consciousness, and some of the attributes of tertiary consciousness, including components of self-awareness and decision making, creativity based on at least one definition, and even an AI form of dreaming. However, both biological systems and their artificial constructs are limited in their capacity to achieve this complex state. This AI version of dreaming and consciousness has significant limitations. There is, clearly, much missing. What is missing is not complexity or computing capability; we can now create AI systems that exceed or are fully analogous to the complexity and processing capacity of biological systems. What is missing are aspects of mind. The short list includes self-reflexive consciousness, significance and meaning, inspiration, innovation, and empathy. The longer list includes aspects of mind that are even more difficult to define: compassion, conscience, transcendence, and ecstasy.

As the mysterium philosophers have indicated, there are many components of dreaming and consciousness that we are unable to logically describe. Such components may prove to be especially hard to artificially create. These are aspects of mind that may be beyond our capacity for technical and empirical analysis (44): they may be across the line.

Notes

1. Breton, A. (1924) *The First Manifesto of Surrealism* Ades, D. and Gale, M. (2007) *Surrealism, in the Oxford Companion to Western Art*, ed. H. Brigstocke. Oxford: Oxford University Press. The first Surrealist's Manifesto was written by André Breton and released to the public in 1924. The document defines Surrealism as psychic automatism in its pure state, by which one proposes to express – verbally, by means of the written word, or in any other manner – the actual functioning of thought.

2. *Early Cinema: Primitives and Pioneers (1895–1910)*. London: bfi Video Publishing. G. A. Smith directed *Let Me Dream Again* in 1900; Edison Manufacturing Company released *The Dream of a Rarebit Fiend* in 1906; and the Lumière Brothers created a series of short takes titled and numbered as *Dreams*, as well as their famous *L'arrivée d'un train en gare de la Ciotat* in 1895, which supposedly excited audiences to run from out of the path of the train *en masse*.

3. Baudry, J. (1986) The apparatus, in *Narrative, Apparatus, Ideology*, ed. P. Rosen. New York: Columbia University Press, pp. 236–318.

4. Freud, S. (1914/1973) Remembering, repeating and working-through, in *The Standard Edition of the Complete Psychological Works*, Vol. V, ed. J. Strachey. London: Hogarth Press, p. 511.

5. Lacan, J. (1979) Desire and the interpretation of desire in Hamlet, in *Literature and Psychoanalysis: The Question of Reading: Otherwise*, ed. S. Felman, Baltimore, MD: Johns-Hopkins University Press, pp. 11–52. It is said, for many of the French philosophers and psychoanalysts, that the true impact of their work can only be approached in person, and perhaps that is why there is more in the literature about Lacan than there is by Lacan.

6. Kaplan, E. A. (ed.) (1990) *Psychoanalysis and Cinema*. New York: Routledge, p. 7.
7. Lacan, J. (1979) The imaginary signifier, in *Four Fundamental Concepts of Psychoanalysis*, trans. A. Sheridan. Harmondsworth: Penguin Books, p. 75.
8. Kaplan, E. A. (1990) *Psychoanalysis and Cinema*.
9. Friedberg, A. (1990) A denial of difference: Theories of cinematic identification, in *Psychoanalysis and Cinema*, ed. E. A. Kaplan. New York: Routledge, p. 36.
10. States, B. (1993) *Dreaming and Storytelling*. Ithaca, NY: Cornell University Press.
11. Cavell, S. (1971) *The World Viewed – Reflections on the Ontology of Film*. New York: Viking Press.
12. Bonitzer, P. (1992) Hitchcockian suspense, in *Everything You Always Wanted to Know About Lacan (But Were Afraid to Ask Hitchcock)*, ed. S. Zizek. London: Verso, p. 17.
13. Zizek, S. (1992) Alfred Hitchcock, or, The form and its historical mediation, in *Everything You Always Wanted to Know About Lacan (But Were Afraid to Ask Hitchcock)*, ed. S. Lizek. Verso: London, pp. 1–14. Zizek, ever the rebel, has had his approach filmed by Sophie Finnes in her recent Zizek documentary, *The Pervert's Guide to Cinema* (2006).
14. Griffith, D. W. (1916) *Intolerance* [film]. Metz, C. (1982) *Psychoanalysis and Cinema*. New York: Macmillan. Jung, B. (2010). *Narrating Violence in Post-9/11 Action Cinema: Terrorist Narratives, Cinematic Narration, and Referentiality*. Wiesbaden: VS Verlag für Sozialwissenschaften, p. 67.
15. Jung, C. (1972) The concept of the collective unconscious, in *The Portable Jung*, ed. J. Campbell and C. G. Jung. New York: Viking, pp. 59–70.
16. Nolan, C. (dir.) (2000) *Memento* [film]. British Board of Film Classification 2000-08-15.
17. Oudart, J. (1978) Cinema and suture. *Screen* 18(4): 35–47.
18. Gabbard, G. and Gabbard, K. (1999) *Psychiatry and the Cinema*. (2nd ed.). Washington, DC: American Psychiatric Press.
19. Dayan, D. (1976) The tutor code of classical cinema, in *Movies and Methods: An Anthology*, ed. B. Nichols. Berkeley, CA: University of California Press, pp. 438–451.
20. Cartwright, R., Bernick, N., Borowitz, G. and Kling, A. (1969) Effects of an erotic movie on the dreams of young men. *Archives of General Psychiatry* 20: 263–271.
21. Ullman, M. and Limmer, C. (eds.) (1998) *The Variety of Dream Experience*. New York: Continuum. Buckley, K. (1996) *Among All Those Dreamers: Essays on Dreaming in Modern Society*. Albany, NY: SUNY Press.
22. Allen, R. (1995) *Projecting Illusion – Film Spectatorship and the Impression of Reality*. Cambridge: Cambridge University Press, p. 115.
23. Cristie, I. (1994) *The Last Machine: Early Cinema and the Birth of the Modern World*. London: British Film Institute.
24. Benson, D. (1994) *The Neurology of Thinking*. New York: Oxford University Press, p. 14. Caudill, M. (1992) *In Our Own Image – Building an Artificial Person*. Oxford: Oxford University Press. Pearl, J. (1988) *Probabilistic Reasoning in Intelligent Systems*. San Mateo, CA: Morgan-Kaufmann.
25. Newell, A. and Simon, H. (1972) *Human Problem Solving*. Englewood Cliffs, NJ: Prentice-Hall.
26. Pagel, J. F. (2008) *The Limits of Dream – A Scientific Exploration of the Mind/Brain Interface*. Oxford: Academic Press/Elsevier, pp. 83–87.
27. Davies, P. (2006) *Towards a Science of Consciousness*, Consciousness Research Abstracts Tucson, AZ, *Journal of Consciousness Studies*, pp. 152–153. Advances in cosmology suggest a link between information, complexity and the age of the universe. This development could remove a fundamental obstacle to strong emergence in nature.
28. Crick, F. and Mitchison, G. (1983) The function of dream sleep. *Nature* 304: 111–114; Crick, F. and Mitchison, G. (1995) REM sleep and neural nets. *Behavioral Brain Research* 69: 147–155.

29. Koch, C. (2012) *Confessions of a Romantic Reductionist*. Cambridge, MA: MIT Press. Koch and Searle have delightfully fought this battle on the somewhat public pages of the *New York Review of Books*. It is mentioned that they are friends who see each other at the same cocktail parties.

30. Searle, J. (1997) *The Mystery of Consciousness*. New York: A New York Review Book. McGinn, C. (2002) *The Making of a Philosopher: My Journey Through Twentieth-Century Philosophy*. New York: HarperCollins.

31. http://en.wikipedia.org/wiki/Numerical_weather_prediction, accessed May 23, 2013.

32. Chorost, M. (2011) *World Wide Mind: The Coming Integration of Humanity, Machines, and the Internet*. New York: Free Press, p. 175.

33. Buckley, K. (2009) *Dreaming and the World's Religions*. New York: New York University Press.

34. Van de Castle, B. (1994) *Our Dreaming Mind*. New York: Ballantine Books, pp. 405–439.

35. Ullman, M., Krippner, S. and Vaughan, A. (1973) *Dream Telepathy*. New York: Macmillan.

36. Kleitman, N. (1987) *Sleep and Wakefulness*. Chicago, IL: University of Chicago Press, p. 125.

37. Arzi, A., Shedlesky, L., Ben-Shaul, M., Nasser, K., Oksenberg, A., Hairston, I. S. and Sobel, N. (2012) Humans can learn new information during sleep. *Nature Neuroscience* 15: 1460–1465.

38. Klatsky, R. and Lederman, S. (1993) Spatial and non-spatial avenues to object recognition by the human haptic system, in *Philosophy and Psychology*, ed. N. Eilan, R. McCarthy and B. Brewer. Oxford: Oxford University Press., pp. 191–205.

39. Llinas, R. and Pare, D. (1991) Of dreaming and wakefulness. *Neuroscience* 44: 521–535.

40. Chorost, M. (2011) *World Wide Mind*, p. 175.

41. Church, G. and Regis, E. (2012) *Regenesis: How Synthetic Biology will Reinvent Nature and Ourselves*. New York: Basic Books, pp. 225–253.

42. Scarry, E. (1995) On vivacity: The difference between daydreaming and imagining-under-authorial-instruction. *Representations* 52: 1–26. Scarry, E. (1999) *Dreaming by the Book*. Princeton, NJ: Princeton University Press.

43. Dennett, D. (1991) *Consciousness Explained*. Boston, MA: Little, Brown & Company, pp. 431–456.

44. McGinn, C. (1982) *The Character of Mind*. Oxford: Oxford University Press, p. 19.

11

Crossing the Line: Dreaming and Ecstasy

An artist might advance specifically to get lost, and to intoxicate himself in dizzying syntaxes, seeking odd intersections of meaning, strange corridors of history, unexpected echoes, unknown humors, or voids of knowledge ... but this quest is risky, full of bottomless fictions and endless architectures and counter-architectures ... at the end, if there is an end, are perhaps only meaningless reverberations. **(Smithson, 1968) (1)**

After almost an entire book describing the biology of dreaming and consciousness, we find ourselves at a point of unclarity and diffuseness in this attempt to study topics that typically evade our empirical and scientific logic. We do not know the basis for consciousness. We do not understand the origin of dreaming. These seem to be extreme statements. Yet in this contentious field, they are the closest that there is to scientific fact. Important and basic components of both dreaming and

Dream Science.
DOI: http://dx.doi.org/10.1016/B978-0-12-404648-1.00011-9

consciousness exist outside our capacity for technical and logical analysis. We do not even know what the questions are that we should ask.

> The image is hypnogogically intense in detail and clarity. White tongue & groove siding covers the exterior walls of the village houses. The roofs are cookie-cutter sharp, cut at 45 degrees if not steeper, covered with dark shingles, lined at the edges by precise metal edging. The houses overlap one another in bewildering geometric detail that from my point of view has an almost two-dimensional aspect.

This image came from a dream I had halfway through writing this book. It has almost unlimited personal associations. There is a white village on top the Greek isle of Karpathos, its image preserved in sunlit detail in a framed photograph on my kitchen mantle. The architecture of the dream village reflects my protestant upbringing, as well as the austere lines of the church depicted in the film *The Tree of Life*. Science in that film is presented as an animated abstraction of existential emptiness, without meaning, purpose, or underlying story. The village dream is also associated with another personal dream – that of a young girl standing at a gate that leads through a circular hedge into another white village on a hill. In that dream the girl opened the gate for me. In waking life, she becomes the girl who I will marry. The two-dimensional overlapping of the dream houses is in form of the white pages or chapters of a book. I see this as a book that I am writing.

> I enter the town and open a door into a small white library. Inside, there are shelves and alcoves set with ladders leaning against the stacks of loosely piled books. In dusk of twilight of the room, I walk up to a wooden desk and look down on a book that lies open there. An intensely green line runs across the pages, jagged, cutting between lines of indecipherable words. Light streams out from that line, illuminating my face and the room. The words of the book are blurred squiggles on gray paper. An intense green light spills out from within the book. From within the dream, I know that it is terribly important.

In my dream, I have discovered something that is ephemeral and magic, yet real. In a state of sudden, intense inspiration and delight, I awake. It is a moment of ecstasy.

This portion of the dream has waking continuity with a winter visit to Iceland to a small white library I visited there, where the original text of the Flateyjarbók saga had been safely hidden through centuries of political turmoil, volcanic eruptions, black death, fires, hurricanes, and economic collapse (2). There, on the tiny, treeless island of Flatey in the icy North Atlantic, the only rendition of the saga of Vinland and Leifur Eiriksson survived. The Northern Lights burn in intensity there, green arcs pulsing and crackling, green dragons racing across the axis of the charcoal blue sky. This dream drops deeply into alternative forms of consciousness.

THE STANDARD APPROACH TO THE ILLOGIC OF DREAM SCIENCE

For centuries, philosophers and scientists have been forced to confront the illogical aspects of dreams. Some have responded by examining their own dreams, in the attempt to apply concrete explanations to the delusional and hallucinatory inflections of the experience. Some have extended this approach and applied waking psychiatric diagnostic categorizations to dreams, in what has become a scientifically acceptable approach (3). But to speculate that mind-based aspects of dreams might have a physiological function or a fact-based explanation is far more suspect. This is a journey toward an empirically loose border. Those who have chosen to study dreaming scientifically as a cognitive state are accused, at times, of crossing the hard line between science and magical thinking.

The world would make more sense if dreams did not exist. Crick and Mitchison have suggested that in order to maintain more efficient waking function, extraneous material (dreams) must be expunged from our mental apparatus during sleep. To maintain optimal performance, data and program cleaning are required routinely in non-biological computer systems. They propose that each night a similar defragmentation process must occur in our brains. Extraneous perceptual waking data are removed and meaningless dreams are produced as a cognitive side-effect. This extraneous information includes dream content that has continuity with our waking perceptual experience. It is this content that is being expunged during dreaming (4).

While some scientists have aligned themselves with this perspective, it is a more commonly shared belief that the structure and function of dreams have a biological basis that can be fully explained. The most simplistic form of this belief is that rapid eye movement sleep (REMS) equals dreaming. More sophisticated forms of monistic belief in a single governing principle postulate that known attributes of biological structure can describe the full phenomenology of the dream state. From this perspective it is only a matter of time before technology proves monistic theory correct. To extend their approach beyond the clear biological correlates, neuroscientists expanded and developed this theory into the activation-synthesis model. McCarley and Hobson postulate that all aspects of mind must be secondary to the underlying biological activity occurring in the brain (5). This theory postulates that a biological basis exists for the psychoanalytic constructs of mind. Activation-synthesis has morphed into current "grand" theories of neuroconsciousness that purport to describe the structure and functioning of the mind and the human psyche (see Preface).

Many neuroscientists and philosophers persist in their belief in the biological basis for mind. Adherents emphasize the capabilities of artificial intelligence (AI) in accomplishing operations that are apparently equivalent to mind-based processes: complex associative memory systems, visual presentation of data, emotional representation, and self-analytic feedback loops, as well as the unexpected and sometimes unexplainable results obtained from composite data analysis. Since there is little current scientific evidence to support the existence of biological correlates for many of the aspects of mind, they argue that in the future, with improved technology, specific biological sites will eventually be found (6). This perspective is another form of denial tied to the belief that dreams have no meaning or importance. It is only a matter of time before technology illuminates the neuroanatomical basis for those components of dreaming that are apparently without biological correlates. Until then, dream-based components of mind can safely be ignored.

It is difficult to scientifically study are poorly defined aspects of mind. There have been attempts to attain clarity by redefining aspects of mind that are difficult to understand as correlates of mind that are easier to study. This approach has been used repeatedly in the study of consciousness. Neuroscientists classically redefine consciousness as the study of attention. Attention is far easier to study and in some forms (e.g., the concentration of processing capacity on delineated topics) clearly within the capacity of AI systems. But redefinition can be used to ignore aspects of the state that are not studied. Redefining consciousness as attention deflects from the inability of current technical systems to uncover the origin and function of many mind-based processes.

Informatics scientists focused on AI define consciousness as the capacity for independent volitional decision making. This has led them into philosophical environs in which photovoltaic cells, complex systems, and much of the world around us have this same capability (7). Searle has caricatured this perspective as a form of "pantheism" in which every object is both conscious and alive. Christof Koch is evidently quite upset to be characterized in such a fashion (8). My perspective is somewhat different. Searle is one of our greatest living philosophers. Koch, a neuroscientist best known for his work with Francis Crick, has gained his attention. Koch may be on to something.

Hunter–gathering tribes most often view their reality as controlled by unseen forces (9). Their waking experience is controlled by outside imbued forces rather than personal trial and error. They live most often in small-scale social structures that function without religious specialists or philosophers (10). Each individual observes, participates in, and integrates this form of magic into the most mundane aspects of their lives.

Hunting is a holy occupation, game animals are spiritual creatures, and the entire way of life is viewed in a holy light:

> Once an old man and his son were very expert in hunting. And the son dreamed that he cohabited with the caribou. It seemed that he killed a great many caribou. Once, when it happened during the winter he said to his father, "I will depart." ... Then he sang: "the caribou walked along well like me. Then I walked as he was walking. Then I took his path. And then I walked like the caribou, my trail looking like a caribou trail where I saw my tracks. And so indeed I will take care of the caribou. I indeed will divide the caribou. I will give them to the people. It will be known to me." ... For so now it is as I have said. I, indeed, am Caribou Man (Ati'k'wape'o). So I am called. *(Nabe'oco, 1923) (11)*

The daily need to hunt and gather creates a world of the eternal present in the culture of those living in this social milieu. If food is not obtained in each present of each day, there is nothing to eat. Eventually, before starvation, the settlement moves following the primary food storage available – the natural changes that occur with season, plant growth, and animal migrations (12). The ancient artists of the caves lived this way in a societal structure similar to that of modern hunter–gatherers (13). Their worldview reflected this waking experience. Their art and dreams reflected and affected their world.

DREAM ILLOGIC IN RELIGION AND PHILOSOPHY

The earliest recorded written records attribute dreams to an outside source. In most cases dreams came from a deity. In taking this approach, ancient authors were admitting to their inability as conscious beings to understand these workings of the mind. Their dreams were given a meaning attributed to something or someone bigger and smarter. This sensible approach had its own empirical limitations then, just as it does today. Almost everyone dreams. Some dreamers are women, some children, and some the criminally insane. Based on most social, political, moral, and religious mores, their dreams should have no importance or meaning. The dreams of righteous adult males, our shamans and leaders, have the potential to be even more destructive, manifesting into religious and political conflict. And there are the nightmares that come to "good" people, even to anointed kings and priests. Should nightmares be interpreted as evidence for the existence of the devil or gods of darkness?

Some historically significant dreams are dreams of religious ecstasy, such as Jacob's dream of the stairway to heaven – a dream that still defines the political and religious limits of our modern world (see Chapter 2). In a cave in the desert, Muhammad, visited by Gabriel in his dreams, wrestled each night with the angel, attempting to understand truth and the divine. In the morning, he would wake and write down the angel's words while still vivid in his memory, "as if they had written a message on my heart" (14). Such ecstatic dreams comprise our holy books and define our religious beliefs. As Ralph Waldo Emerson has pointed out, an ecstatic world, available in dreams and in episodic dream-like waking, can lie beneath even our rational and scientific endeavors:

> Underneath the inharmonious and trivial particulars [of our daily lives], is a musical perfection, the Ideal journeying always with us, the heaven without rent or seam ... By persisting to read and think, this region gives further sign of itself, as it were flashes of light, in sudden discoveries of its profound beauty and repose, as if the clouds that covered it parted at intervals, and showed the approaching traveler the inland mountains, with the tranquil eternal meadows spread at their base, whereon flocks graze, and shepherds pipe and dance. *(Emerson, 1844/2001) (15)*

Some religious traditions utilize a personally empirical approach that includes meditation. Meditation is an approach that can be used in an attempt to understand interior aspects of consciousness. A multitude of techniques, focused and unfocused approaches, can be used interior or exterior to religious traditions and training. Some approaches focus on techniques and levels of attained experience, and others on ritual and/or prolonged experience in the meditative state and avoid attempts at self-analysis. Meditation can lead to a change in the perception of the exterior world in which the accomplished meditator becomes able to distinguish between the illusions of "deluded consciousness" and an awareness of the essence of subject and object in a varying conscious universe (16). To the experienced meditator, the exterior world can seem more "real" and apparent. A characteristic shared by the acolytes of most meditative traditions is a striving to attain a state of enlightenment. That state, itself poorly defined, may be somewhat like ecstasy.

The attempt to differentiate truth from illusion, real dreams from false, has led to many aspects of philosophy, including the scientific method (see Chapter 3). Some philosophers have concluded that aspects of mind, including dreams, may be beyond our human capacity to understand. The mysterium philosophers Nagel and McGinn are quite comfortable with the argument that our processes of objective logic may not apply to the study of aspects of the mind such as dreaming and consciousness. Mind is potentially on the other side of a cognitive interface that we cannot cross using any technology that is based on waking perceptual tools and logic (17).

CROSSING THE LINE

Both scientific and philosophical/religious approaches to dreaming tend to arrive at similar perspectives when freed from belief systems that preclude evidence. Scientifically, we find ourselves unable to explain the origin and/or basis for either dreaming or consciousness. There are philosophical as well as religious arguments suggesting the same. Consciousness is indefinable, unexplainable, and perhaps unknowable, and dreams are among the most difficult forms of consciousness to study and understand. Yet we all experience consciousness and we almost all experience dreaming. These aspects of mind are central to our existence. Even when we deny or ignore them, they will not go away.

Fortunately, there are artists. Scientists, philosophers, and theologians often trace artists' paths, their journey into the nexus of understanding of external and internal realities. The same diffusely defined cognitive states, so difficult to empirically explain, form the meat-and-potatoes of artistic work.

Artists use tools. These tools can be deceptively simple: a brush, a pen, a vibrating wire. They can be complex: a digital camera, an orchestra, a computer program. In an artist's hands, the tool comes to life, almost conscious in interface. The artist manipulates the tool to facilitate expression and passion outside its intended function. Most analysis of art concentrates on the completed process, the final version of a painting, a song, or the final cut of a film. The aspects of creation that occur in the process are often beyond the analysis that comes in the ascribing of theories and philosophy to that art (18). It is in this creative process that dreams are both used as tools and incorporated into the invented artwork. In concentrating on the process rather than the product, the functional role that dreaming plays in creativity becomes obvious (see Chapter 1). Many artists incorporate subjectively unstable reflexive aspects of dreams into their creative process. This can lead to artistic outcomes that are beyond the conscious expectations of the artist. This takes place in film-making, a very public and shared experience of creation.

In film-making, a director can use the camera to intentionally "cross the line," a term that is almost unexplainable outside being a witness on set. The camera operator uses the movie-camera to record scenes, most often following a planned and scripted storyboard for the film. It is important to maintain continuity between shots and between scenes. Continuity can be as simple as matching clothes from scene to scene, but it also applies to the continuity of eyeline, shot to shot. The actors' eyeline must match the eyeline of person they are speaking to or the object they are observing; otherwise the audience will find itself removed from the "willing suspension of disbelief" that characterizes the movie-watching experience. This is so crucial to the audience's acceptance of the created world as presented in the film that there is a crew position assigned to monitor it, shot to shot. A "script supervisor" maintains continuity between shots and scenes, related to props, dialogue, wardrobe, and eyeline. The camera operator, filming without assistance in maintaining eyeline continuity, could easily drift to more allusive qualities in following story movement, reflective shadows, or a character's contemplative shift of gaze. Pushing the camera's focal point to cross this sometimes intangible line on set creates a misconnect between shots, becoming a much larger problem for the editor, trying to fit shot-to-shot puzzle pieces together to create a scene that the audience can visually read. Despite these, and other concrete attempts to create apparent reality in a film, the complexity of the collaborative film-making process ensures that no scene will be filmed exactly as planned. Metaphorically, most great films will cross the line in some way. Camera shots will be from impossible angles, words will simmer in wild sound from off-camera, focus will wander, actors will reset the pace in skipping a beat, or maybe two, creating a reflexive instability such as in life. This is the art of directing. A director takes these unvaryingly imperfect scenes and edits them together to create a filmatic experience that can be more real and impactful than an experienced dream. In this way, the film-maker frees the film from the immediate and concrete demands of continuity and narrative (19). The best scenes, by the greatest film-makers, are those that allow the accidents and imperfections of both the camera and the process to create the unexpected. The camera, the brush, the pen, set free, become almost uncontrollable creatures, needing only the loosest of guidance in finding their own path, creations of an almost conscious world aligned with the mind of their creator.

Writers consciously use techniques that in effect cross the line to create such dream-like virtual realities. Elaine Scarry has studied the techniques that novelists and poets use to create imagined states. From Homer forward, she has found a consistent use of a set group of descriptive techniques. These include:

- the use of material antecedents of perception (the structure, texture, and play of light upon an object)
- emphasis on the faintness, two-dimensionality, and fleetingness of mental imagery
- presentation of the weightless and translucent aspects of images relative to each other.

These techniques can lead the reader to an experience of sensory mimesis in which the experience derived by reading becomes more like direct perception than imagined imagery (20). These descriptions of imagined reality are presented by a character in the narrative as if they were directly experienced. The writer, in producing an imagined state for the character, has a far different effect on the reader. A relationship, a correspondence, develops between the world of the reader and the world contained within the book, and the reading experience achieves an intensity and vivacity greater than that of waking experience (21). The story expands, offering multiple time frames as well as the distortion of multiple voices, perspectives, and memories. The story becomes not merely about things that have happened, but about events that can change us (22).

Scarry has also looked at the formal characteristics that contribute to the vivacity of moving pictures. These same concepts can be applied loosely to film. She calls the first of these characteristics "radiant ignition," and relates this example from the ancient *Iliad*:

> She flung to the winds her glittering headdress,
> The cap and the coronet, braided band and veil,
> All the regalia golden Aphrodite gave her once,
> The day Hector, helmet aflash in sunlight,
> Led her home to Troy from her father's house
> With countless wedding gifts to win her heart. (*Homer's* Iliad) (23)

This passage includes attributes that induce a sense of motion as well as the experience of sensory mimesis: rarity (the headdress floating on the winds), authorial instructions for the mental movement of the image, a nature basis for the image, the motion of circling or returning, and the visual operative move of repicturing. The authorial directions suggest that the reader/viewer create a mental composition. Aspects of motion incorporated into writing can be used to "dislocate" both the story and the reader into a state that is "in between," inducing the experience of an imaginary and dream-like landscape (24). Cormac McCarthy's Pretty Horse dream (see Chapter 8) leads the reader into such a moving geographic dreamscape (25). The vivacity and interiorly experienced reality induced by this moving and translucent passage is far greater than that which could be derived from a direct narrative description. Such techniques, first developed long ago by Homer, are used repetitively by today's authors and film-makers.

From film and from fictional literature, this is artistic evidence for the ability of dreaming to induce ecstasy. It is through entering and then awakening from a dream or a dream-like state that we are more likely to experience the sudden intense experience of waking illumination. These experiences are seemingly more present and more real than the perceptual waking attention that we focus on our surrounding environment. Artists and writers experience this ecstasy in their work: in the midst of their process, they may lose their orientation to place and time, immersed in work that is often described by the artist as dream-like (26). These experiences of the imagination, often derived and adapted from dreams, are likely to be required for the creative process (27).

These imaginative experiences are often far more real for the individual than any empirically described correlates. A dream can be viewed as its associated neuroanatomy, neurochemistry, or electrophysiology. A dream can be understood as different forms of consciousness that occur during

sleep and in the transition to wakefulness. A dream can be viewed as constituent memories, images, and emotions from sleep that we structure
into a narrative story. These descriptions of dreaming are scientifically
and empirically correct, and are very useful for both the scientist and the
dreamer as constructs that can be used to understand how and why we
dream. Yet none of this elegant scientific structure defines the essence of
the experience of dreaming. None of the reproducible scientific evidence
and logic addresses the potential ecstasy of such an experience, and how
such an experience may affect our waking thoughts and functioning.

From where I sit, I can look out into a landscape of trees, some
browned with beetle kill. Billowing clouds form over high peaks still
snowcapped from the late snows of a heavy winter. They form dark
slowly moving shadows across the landscape. I sit on a pink granite
boulder that is streaked with veins of white quartz. On the hillside, the
rotting trunks of trees lie on the ground next to the gray stumps from
which they had been cut over 100 years ago. The air is crystal clear, and
the light at midday is intense, necessitating dark glasses. The needles
under the pines form concentric piles and patterns where the jays and
nutcrackers have used them for winter storage. A large ant crawls across
my shoe. I flick it away.

From where I sit, I look up at the needles of the spruce that, unnoticed
by me, has grown next to this rock on the hillside. Its needles this spring
are intensely green. Perhaps they are no greener than they have ever
been. But today, the day after my dream of Iceland, they are the greenest
green that I have ever seen.

THE ECSTASY OF CAVE PAINTERS

As art, we categorize the cave paintings in a world of their own, with
connections and similar patterns to folk and intuitive art. The study of
their artistic process is less encumbered by the language of modern and
post-modern technical analysis. We are less inclined to discuss point of
view, the eye as camera, psychoanalytic theory, or school of training.
Archaeologists and anthropologists agree that the creation of these paintings reflects more than a simple depictive skill. These paintings reflect
the ecstatic. Their argument uses psychoanthropological evidence to
conclude that the origin of that ecstasy was more than likely the drug-
induced shamanistic trance (see Chapter 2). Evidence to the contrary
from the fields of dream science is addressed in that same chapter. The
ecstatic component of cave art was far more likely to have come from
dreams – the same easily available source that artists utilize today.

Dream consciousness has similarities with the worldview of hunter–
gatherers: a survival-induced present, an environment in which the

smallest and most mundane of experiences had significance and meaning. The cave painters lived, functioned, and created in such a world. They painted the images, symbols of food, predators, beasts in an eternal present coming and going in a translucent and transcendent universe. Cave artists painted images that they visualized; the outlines of primitives and archetypes from any perspective, in varying light, are intrinsic representations of cave bears, horses, rhinos, ibex, and mastodons. These are not individualistic representations and they have little in common with drug-induced hallucinations. These representations were used to present creatures in the same manner for tens of thousands of years. These images are art requiring cognitive capabilities that would distinguish our species. They were shared and used by their tribal community, and they can still be shared and understood. These are the images of dreams.

Dream consciousness and the Paleolithic worldview resemble the state of pantheism described by Koch in which objects and actions have their own forms of consciousness. Despite philosophical objections, this is not an unscientific view. For youth, the realization that much of the surrounding environment is alive and making independent self-motivated decisions comes as a remarkable and amazing realization. It becomes even more amazing when coupled with the scientific knowledge that this exterior life shares characteristics and consistent patterns of behavior. When we look outside ourselves, so much of our surrounding environment, our exterior life, is alive. So much really is conscious.

The majority of the non-human organisms in our environment possess a consciousness that is quite different from our own. For most, it is primary and reflexive of the surrounding environment, possessing limited self-awareness. Some organisms, the domesticated pet being one, can integrate perceptual and sensory input using second-level cognitive processing: thoughts, emotions, the sense of self, learning, mental imagery, and orientation. But it is humans that integrate primary and secondary cognition into processes of tertiary (executive-level) consciousness to allow the function and formation of a society. Humans organize their thoughts independently, operating from within a state of reflexive consciousness on a timeline that includes the personal, the past, and the future. These capabilities provide us with an awareness of our own thoughts and an understanding of the behaviors of others. Dreaming is a type of reflexive consciousness that allows us to self-consider and interface with the otherwise inaccessible forms of consciousness present during sleep.

Dreaming is part of a reflexive feedback system most active during sleep. It is composed of visual, emotional, and memory content that has continuity with our waking experience. In our dreams, this content exists and morphs in forms of consciousness that are outside our volitional control. These remarkable forms of consciousness are structured with basic forms of narrative and grammar. On waking, we organize these experiences into narrative form. These dreams are open to creative associations and free-ranging interpretation. These narratives form the essence of our stories. Those images, reflective of our dreams, use dream-like forms to bring us into imagined stories, from where we are able to re-experience our own reality with an added vivacity, wonder, and amazement.

Artists can bring the world that exists across the line into our perceptual waking world. This is accomplished not using the logic of brain-based process, but rather using descriptive and representational constructs. These patterns of representation are present in this world as an aesthetic interplay between the artist and the viewer/reader. This aesthetic interplay, this art, is, more than almost anything else, an expression of ecstasy.

DREAM SCIENCE

The scientific method has proven to be a powerful tool even when used to examine the unusual cognitive experience of dreaming. Descartes discovered this method (ecstatically in a shower of sparks) during dreaming. It is a shame that it has rarely been applied in essential philosophical and evidence-based form to the study of its state of origin. Dream theoreticians of the past 110 years pushed the boundary of scientific method when they dropped the requirement for evidence. Many became shamans – believers in their own dreams. From the perspective of dream science, neuroconsciousness theories have functioned in a quite negative way to suggest that dreaming is a simplistic and deluded cognitive state comprised of degraded mentation.

The scientific evidence suggests something quite different. To paraphrase Shakespeare, "we sleep, perchance to dream" (28). Dreaming is likely to be a primary function of sleep. We utilize forms of waking that have dream-like characteristics in order to expand our capacities of waking functioning, and in order to incorporate our dreaming consciousness into waking experience. Viewed independently of the proposed biological correlates, dreams are special forms of consciousness, functioning as an interface with cognitive processing occurring in sleep. They are multifaceted and alternative forms to waking, exemplifying the tertiary process of self-reflective consciousness. Changing our understanding of dreams, we change our understanding of every one of the areas of science and philosophy that have ever chosen to consider a potential role for dreaming in their field. Consider the wide focus of this book. Consider dreams free of preconceptions, and in each field addressed, whether archaeology, anthropology, religion, philosophy, electrophysiol-ogy, neurochemistry, neuroanatomy, linguistics, narrative, vision, story, film, literature, or art; in each case, a reconsideration of dreaming leads to a sometimes profound change in the understanding of that field of study. But any change in true understanding extends far beyond changes in our fields of knowledge.

Each of us uses dreams personally. We use our dreams in the process-ing of waking experience. We use dreams in our creative process, as tools for crossing the line to produce new and alternative approaches to the sometimes difficult processes of life. We can use our dreams to create art, literature, and even science. We use our dreams to help in understanding ourselves, our potential, our stories, and our meaning in this existence. To be human is to dream.

Notes

1. Smithson, R. (1968) A museum of language in the vicinity of art. *Art International* March: 21; Lippard, L. R. (1973/2001) *Six Years: The Dematerialization of the Art Object from 1966 to 1972*. Berkeley, CA: University of California Press, p. 44.
2. Rowe, E. A. (2005) *The Development of Flateyjarbók*. Odense: University Press of Southern Denmark.
3. Hobson, J. (1999) *Dreaming as Delirium*. Cambridge, MA: MIT Press.
4. Crick, F. and Mitchison, G. (1983) The function of dream sleep. *Nature* 304: 111–114; Crick, F. and Mitchison, G. (1995) REM sleep and neural nets. *Behavioral Brain Research* 69: 147–155.
5. McCarley, R. and Hobson, J. (1975) Neuronal excitability modulation over the sleep cycle: A structural and mathematical model. *Science* 189: 58–60.
6. Dennett, D. (1991) *Consciousness Explained*. Boston, MA: Little, Brown & Company, pp. 431–456; Penrose, R. (1990) *The Emperor's New Mind*. New York: Oxford University Press; Churchland, P. (1986) *Neurophilosophy: Toward a Unified Science of the Mind/Brain*. Cambridge, MA: MIT Press, p. 272.
7. Koch, C. (2012) *Confessions of a Romantic Reductionist*. Cambridge, MA: MIT Press.
8. Searle, J. (2013) Can information theory explain consciousness? *New York Review of Books*. January 10, pp. 54–58.
9. Whitley, D. S. (2009) *Cave Paintings and the Human Spirit: The Origin of Creativity and Belief*. Amherst, NY: Prometheus Books.
10. Kew, J. (1976) Foreword to the new edition, in *Naskapi: The Savage Hunters of the Labrador Peninsula* F. Speck (1935/1977). Norman, OK: University of Oklahoma Press, p. xii.
11. Nabe'oco (1923) *Naskapi: The Savage Hunters of the Labrador Peninsula*, transcribed by F. Speck (1935/1977). Norman, OK: University of Oklahoma Press, p. 81.
12. Lewis-Williams, D. and Pearce, D. (2005) *Inside the Neolithic Mind*. London: Thames & Hudson, p. 20.
13. Lewis-Williams, J. (2002) *The Mind in the Cave: Consciousness and the Origins of Art*. London: Thames and Hudson; Curtis, G. (2006) *The Cave Painters: Probing the Mysteries of the World's Finest Artists*. New York: Anchor Books.
14. *Qur'an* (1974), trans. N. J. Dalwood.
15. Emerson, R. W. (1844/2001) Experience, in *Emerson's Prose and Poetry*, ed. J. Porte and S. Morrin. New York: Norton, pp. 207–208.
16. Wallace, B. (2012) *Dreaming Yourself Awake: Lucid Dreaming and Tibetan Dream Yoga for Insight and Transformation*. Boston, MA: Shambhala, pp. 135–150.

17. McGinn, C. (1991) *The Problem of Consciousness*. Oxford: Blackwell; McGinn, C. (2002) *The Making of a Philosopher: My Journey Through Twentieth-Century Philosophy*. New York: HarperCollins; Nagel, T. (1974) *Mortal Questions*. Cambridge: Cambridge University Press, p. 47.
18. Eastwood, S. (2009) The film is in front of us, in *Telling Stories: Countering Narrative Art, Theory and Film*, ed. J. Tormey and G. Whitley. Newcastle upon Tyne: Cambridge Scholars Publishing, pp. 253–265. It is difficult to address the creative process without experience in that process. My experience is in film-making and as such the dream-like and non-dream-like aspects of that experience seem quite obvious.
19. Astruc, A. (1948) The birth of a new avant-garde: La camera stylo, quoted in Armes, R., *French Cinema Since 1946*, Vol. 2, *The Personal Style*. London: Zwemmer, pp. 137–144.
20. Scarry, E. (1995) On vivacity: The difference between daydreaming and imagining-under-authorial-instruction. *Representations* 52: 1–26; Scarry, E. (1999) *Dreaming by the Book*. Princeton, NJ: Princeton University Press.
21. Jacobus, M. (1999) *Psychoanalysis and the Scene of Reading*. Oxford: Oxford University Press, p. 18.
22. O'Neill, M. (2009) Introduction: Theories and criticism, in *Telling Stories: Countering Narrative Art, Theory and Film*, ed. J. Tormey and G. Whitley. Newcastle upon Tyne: Cambridge Scholars Publishing, pp. 2–3.
23. Homer (1990) *Iliad*, trans. R. Fagles. New York: Benard Knox, pp. 550–555.
24. Martin, C. (2009) The methodology of mailmen: On delivering theory, in *Telling Stories: Countering Narrative Art, Theory and Film*, ed. J. Tormey and G. Whitley. Newcastle upon Tyne: Cambridge Scholars Publishing, pp. 65–74.
25. Scarry, E. (1999) *Dreaming by the Book*, pp. 239–248.
26. Pagel, J. F. (2008) *The Limits of Dream – A Scientific Exploration of the Mind/Brain Interface*. Oxford: Academic Press/Elsevier, pp. 157–160.
27. McGinn, C. (2004) *Mindsight: Image, Dream, Meaning*. Cambridge, MA: Harvard University Press, pp. 151, 159.
28. Shakespeare, W. (1604/1975) Hamlet, Act 3, Scene 1, *The Complete Works of William Shakespeare*. New York: Gramercy Books, p. 1088.

Index

Printed and bound by CPI Group (UK) Ltd, Croydon, CR0 4YY

03/10/2024

01040428-0002